Annual Meeting of the American Electrophoresis Society 2015 (AES)

Topical Conference at the 2015 AIChE Annual Meeting

Salt Lake City, Utah, USA
8-13 November 2015

ISBN: 978-1-5108-1843-9

Printed from e-media with permission by:

Curran Associates, Inc.
57 Morehouse Lane
Red Hook, NY 12571

CURRAN ASSOCIATES INC.
proceedings
.com

Some format issues inherent in the e-media version may also appear in this print version.

Printed by Curran Associates, Inc. (2016)

AIChE
120 Wall Street, FL 23
New York, NY 10005-4020

Phone: (800) 242-4363
Fax: (203) 775-5177

www.aiche.org

Additional copies of this publication are available from:

Curran Associates, Inc.
57 Morehouse Lane
Red Hook, NY 12571 USA
Phone: 845-758-0400
Fax: 845-758-2634
Email: curran@proceedings.com
Web: www.proceedings.com

TABLE OF CONTENTS

422958 Characterization of an Internal Standard for Phosphotyrosine Western Blotting

Abstract:

Characterization of an Internal Standard for Phosphotyrosine Western Blotting

Western blotting in combination with carrier ampholine 2D gel electrophoresis (CA-2D) and enhanced chemiluminescence detection (ECL) is the most sensitive method known for detecting specific proteins in complex mixtures. Our lab has been using an anti-phosphotyrosine antibody in combination with CA-2D western blotting to visualize specific receptor tyrosine kinases (RTK) in human tumor samples. Active RTKs, which are cancer drivers, have multiple phosphotyrosines (pTyr). The same RTK without pTyr is inactive, EGFR for example. One problem we have encountered however, is in quantifying results. The time course of ECL light emission varies dramatically over the first hour. Thus the same exposure time, 3 min for example, gives a different visual pattern at the beginning of the hour than at the end for the same sample. One way of quantifying the western blot reaction is to include an internal standard. Results from the spot density of the protein of interest are then normalized by expression as a ratio of the spot density of the standard. Surprisingly few pTyr standards are available. In this talk we will discuss results from three potential pTyr internal standards: JNK1 protein from the University of Dundee, phosphotyrosinylated BSA from Sigma, and phosphotyrosinylated ALK fragment from ProQuinase.

EMSL

Pacific Northwest National Laboratory | U.S. Department of Energy

New strategies for preconcentration, CE injection, and ionization for mass spectrometry-coupled microfluidic analyses

Ryan Kelly
Pacific Northwest National Laboratory
Richland, Washington, USA

Scientific Innovation Through Integration ■ www.emsl.pnnl.gov

Proteomics

- Comprehensive approach for characterizing proteins in biological samples
- Promising for early diagnosis of disease
- Thousands of proteins from a ~microgram sample

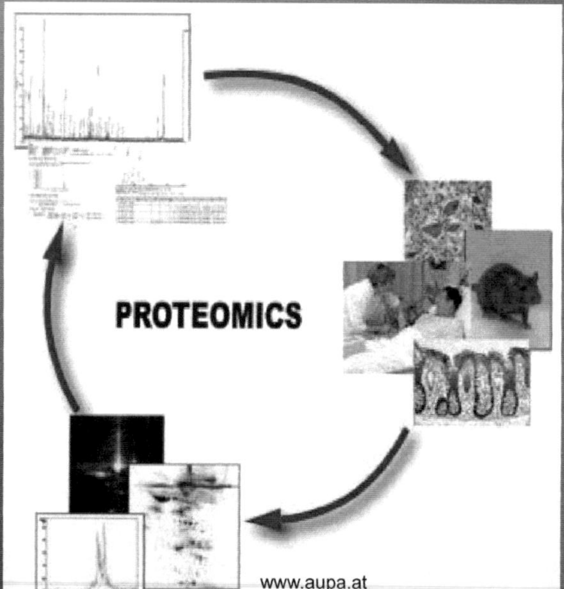

EMSL

Steps of a Proteomic Analysis

- Sample preparation
  - Sensitivity
- Liquid-Phase Separation
  - Sensitivity
  - Throughput
  - Peak capacity
- Ionization
  - Sensitivity
- Mass spectrometry
  - Sensitivity
  - Mass measurement accuracy
  - Resolution

- As sensitivity increases:
  - Increased proteome coverage
  - Smaller sample sizes
  - Single cells?

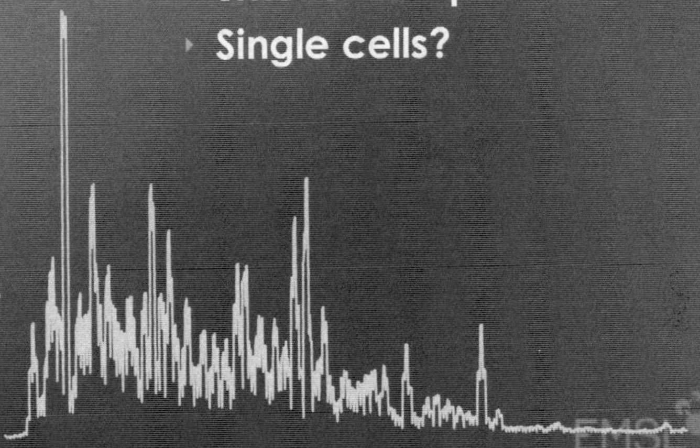

EMSL

Why Peak Capacity?

- Less complex mixtures delivered to MS
- Reduced ionization suppression
- Increased dynamic range
- Greater proteome coverage

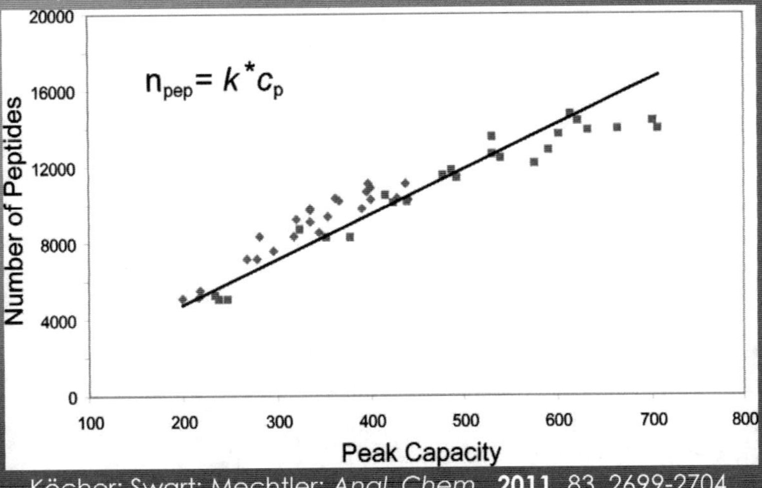

Köcher; Swart; Mechtler; *Anal. Chem.* **2011**, 83, 2699-2704.

EMSL

Proteomic Separations: LC-MS and CE-MS

- Liquid Chromatography
  - Larger sample loading capacity (μg vs. ng)
  - Inherent analyte preconcentration
- Capillary Electrophoresis
  - Peak capacity comparable to LC
  - Enhanced coverage of low MW peptides and high MW proteins
  - Improved mass sensitivity (nano-ESI flow rates)
  - Potentially faster separations

EMSL

Existing Approaches for LC-CE

- LC and CE are highly orthogonal
- Solvent gradient from LC is OK for CE
- **Most of sample is wasted**

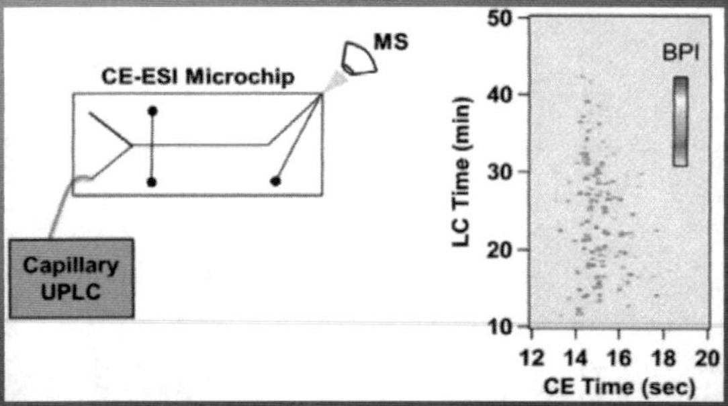

Mellors et al., *AC*, 2013, 4100

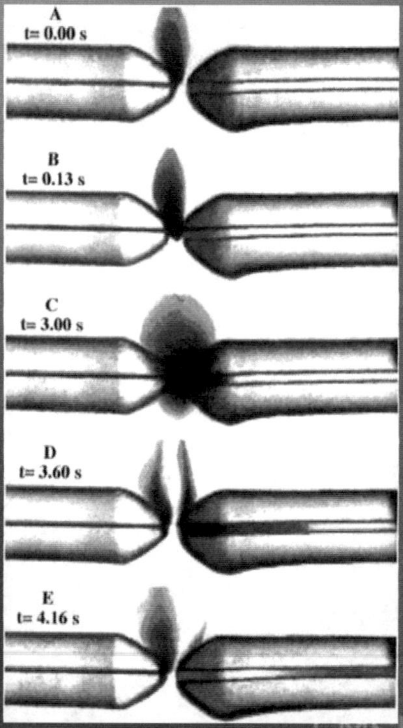

Hooker and Jorgenson, AC, 1997, 4134

EMSL

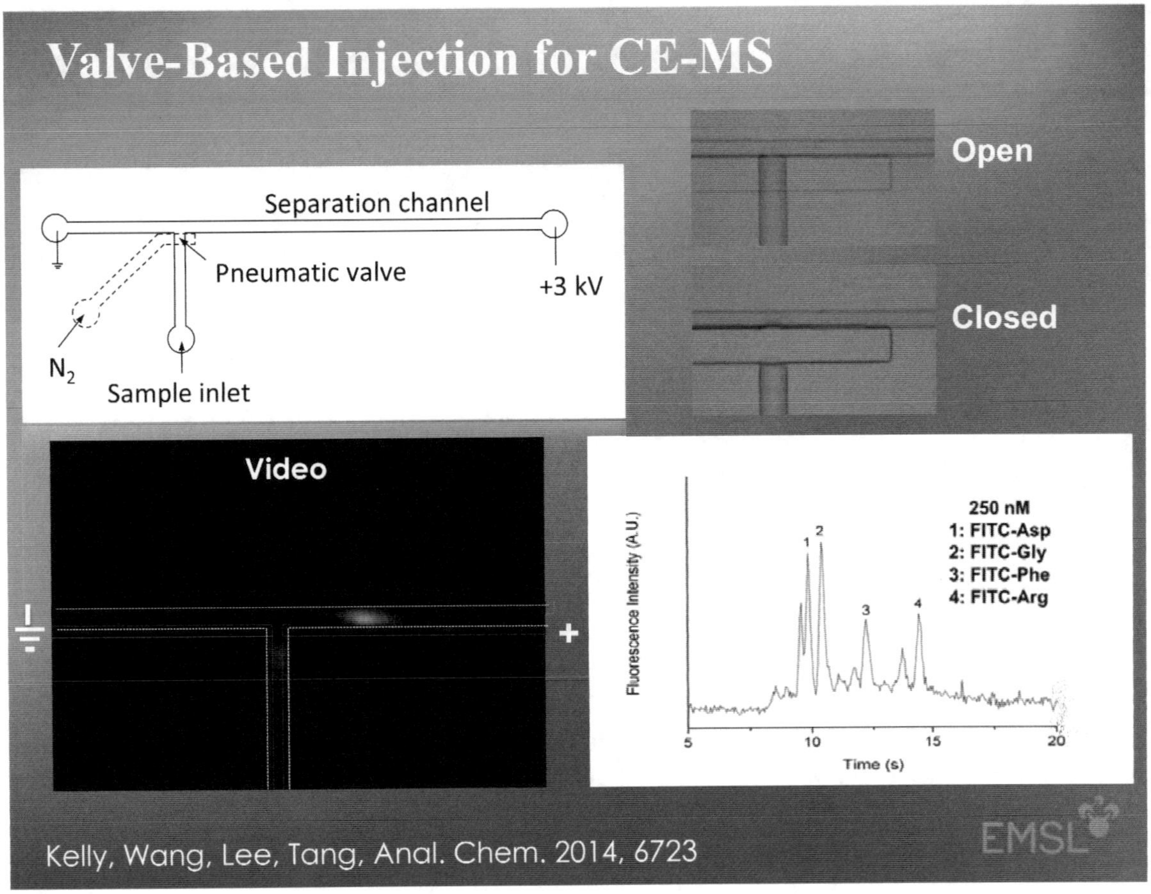

Valve-Based Injection for CE-MS
Separation channel
Pneumatic valve
+3 kV
N_2
Sample inlet
Open
Closed
Video
+
250 nM
1: FITC-Asp
2: FITC-Gly
3: FITC-Phe
4: FITC-Arg
Fluorescence Intensity (A.U.)
Time (s)
Kelly, Wang, Lee, Tang, Anal. Chem. 2014, 6723
EMSL

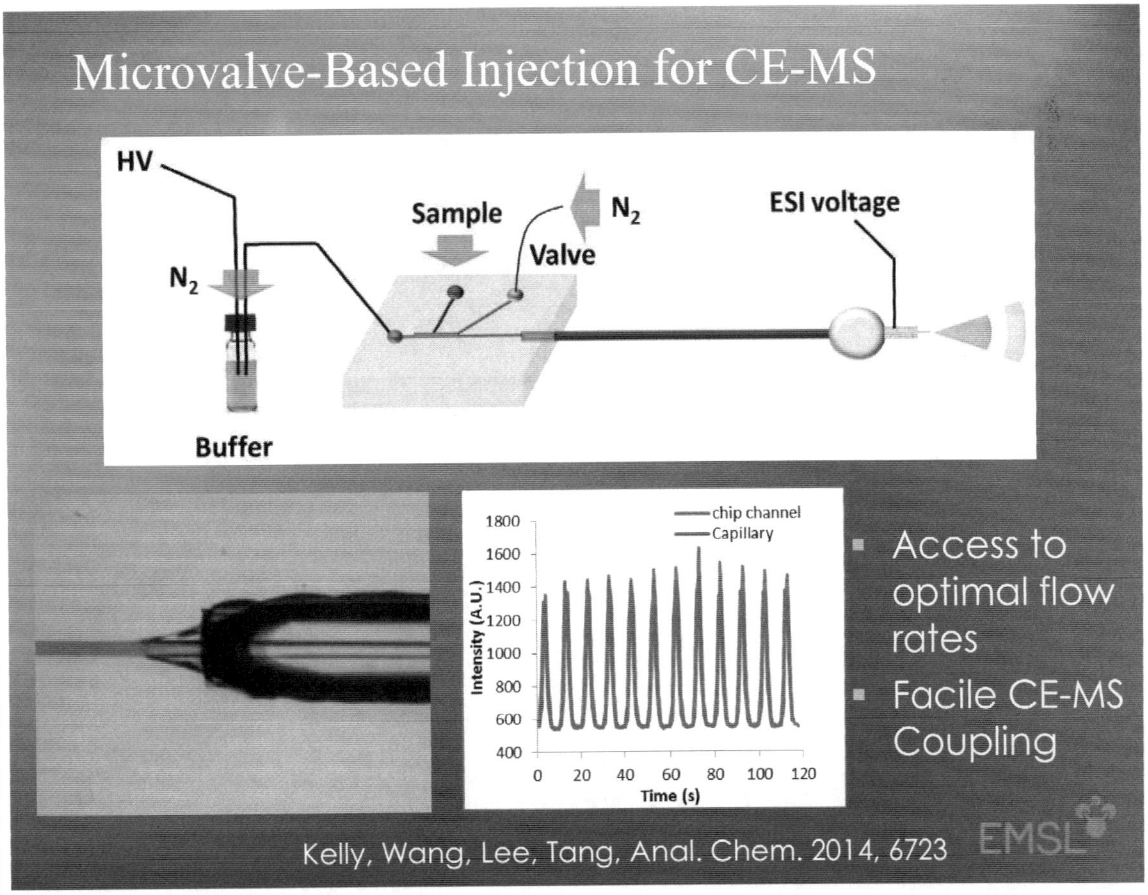

Microvalve-Based Injection for CE-MS
HV
N_2
Sample
N_2
Valve
ESI voltage
Buffer
chip channel
Capillary
Intensity (A.U.)
Time (s)
Access to optimal flow rates
Facile CE-MS Coupling
Kelly, Wang, Lee, Tang, Anal. Chem. 2014, 6723
EMSL

Variable Injection Times

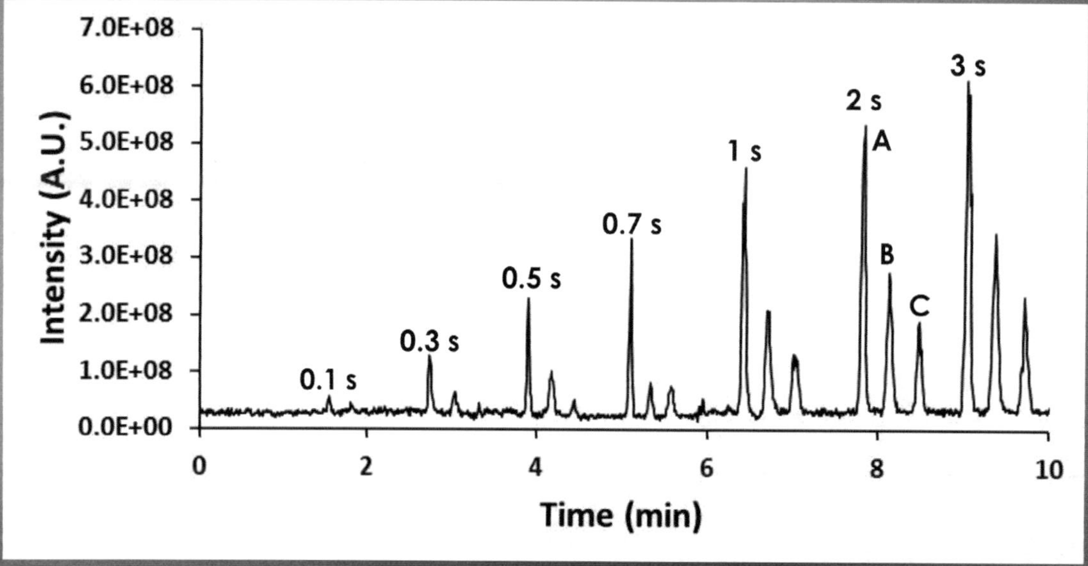

2 psi elution pressure, 20 cm capillary, 30 μm i.d., 13 kV, 10 μM peptides

A: Kemptide, B: Angiotensin II; C: Leucine Enkephalin

Kelly, Wang, Lee, Tang, Anal. Chem. 2014, 6723

Preconcentration for CE

- **Non-electrophoretic methods**
  - **Solid phase extraction**
  - **Immunoaffinity**
- **Electrophoretic methods**
  - **Field-amplified stacking**
  - **Moving boundary stacking**
  - **Isotachophoresis**
  - **Focusing (IEF, EFGF, etc.)**
  - **Nano-porous membrane/nanochannel preconcentration**
- **On line preconcentration and separation: injection issue**

nanochannel

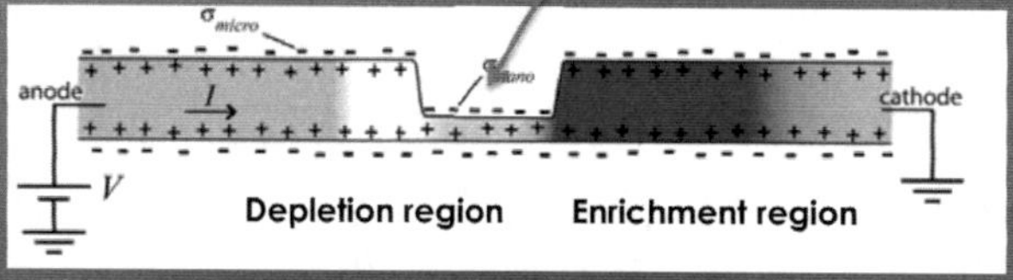

Zangle, T., et al. Anal. Chem. 2010, 82, 3114-3117

Our approach: a microvalve serves as both nanochannel preconcentrator and hydrodynamic injector on a PDMS chip

Schematic of Online Preconcentration and Injection of Anions

1. Setup (same as before)
2. Sample Preconcentration
3. Sample Injection
4. Separation

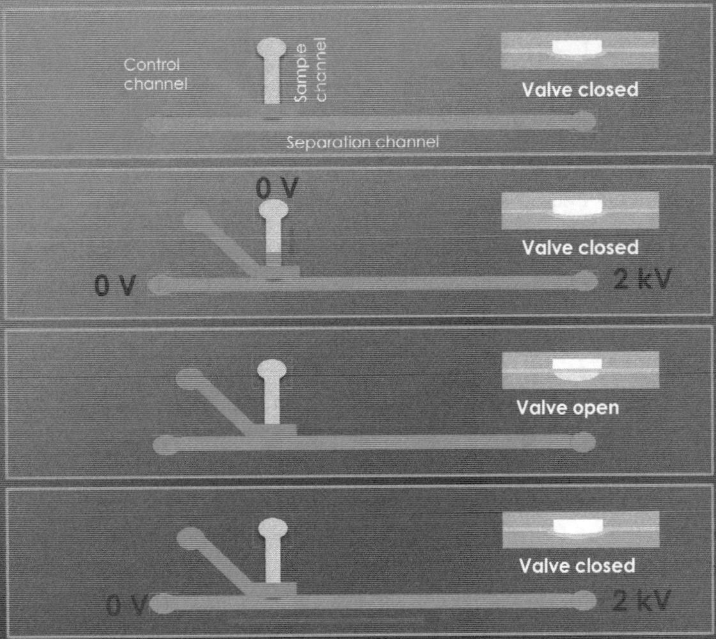

EMSL

Membrane-Based Preconcentration

- Potential applied across closed membrane
- Injection upon valve release
- Up to 1000-fold enrichment
- Potential for lossless LC-CE interface

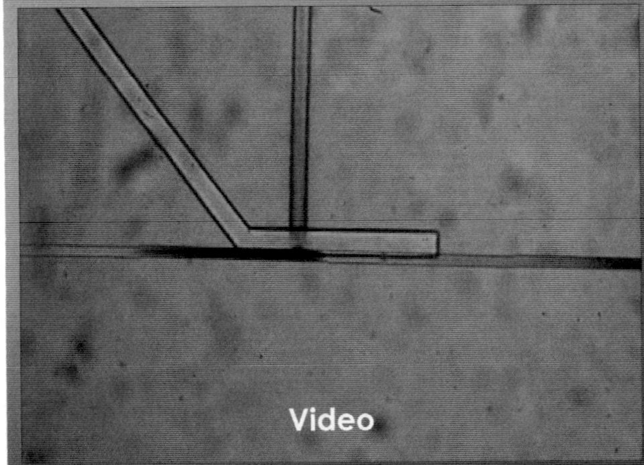

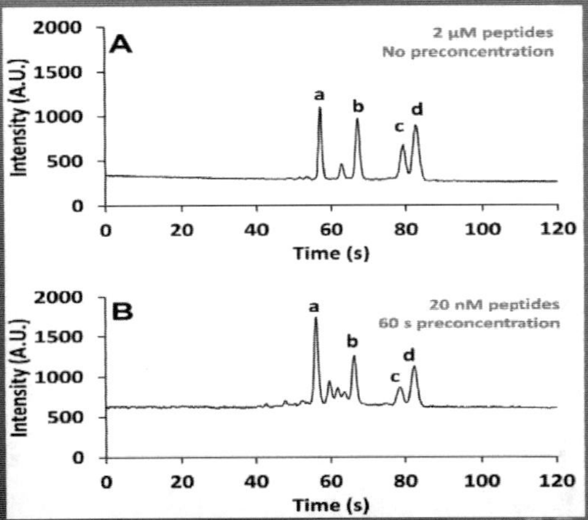

Electrophoresis DOI: 10.1002/elps.201500286

EMSL

Possible Online Coupling LC to CE

- Capture analyte from LC eluent as solvent passes to waste
- Inject analyte periodically for CE separation
- Overcoming hydrodynamic forces is challenging

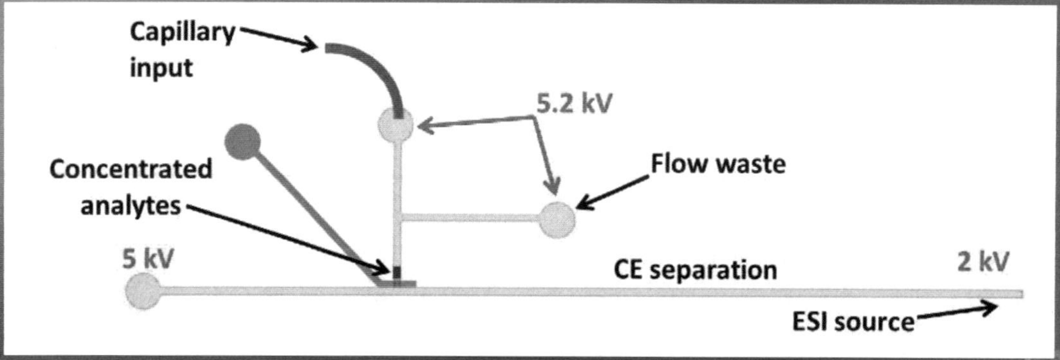

EMSL

Offline Coupling Strategy: Sequential Operations Droplet Array

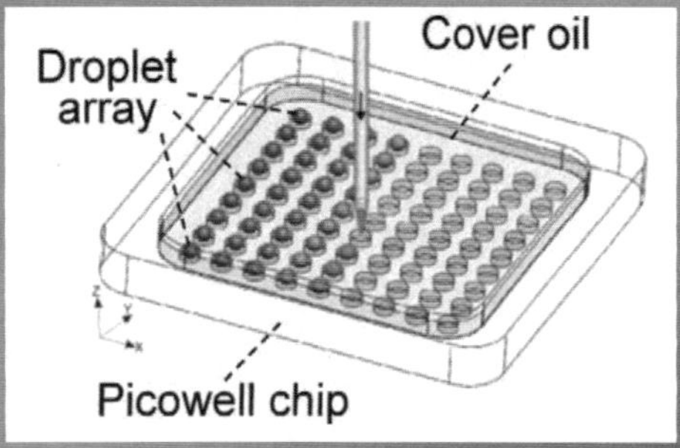

Anal. Chem. 2013, 85, 6723–6731

- Deposit from LC into ~100 nL nanowell arrays
- Preconcentrate from arrays for CE injection

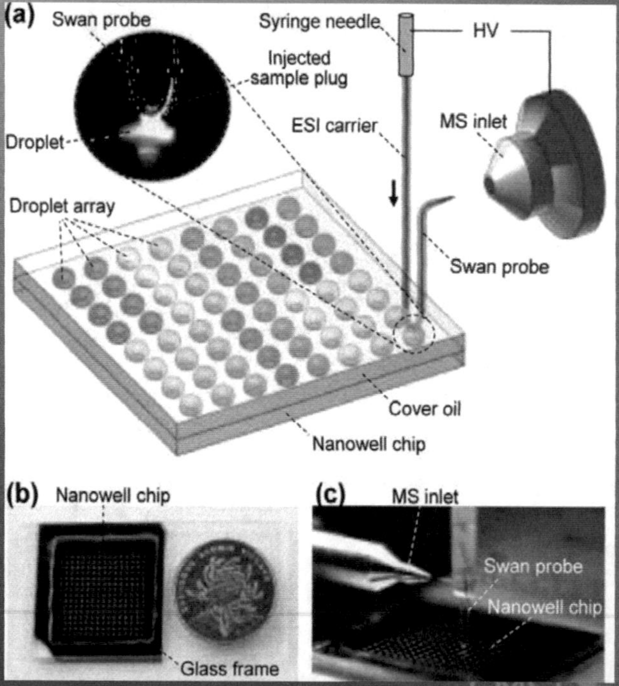

Anal. Chem. 2014, 86, 10796–10803

Adaptation of "Swan Probe"

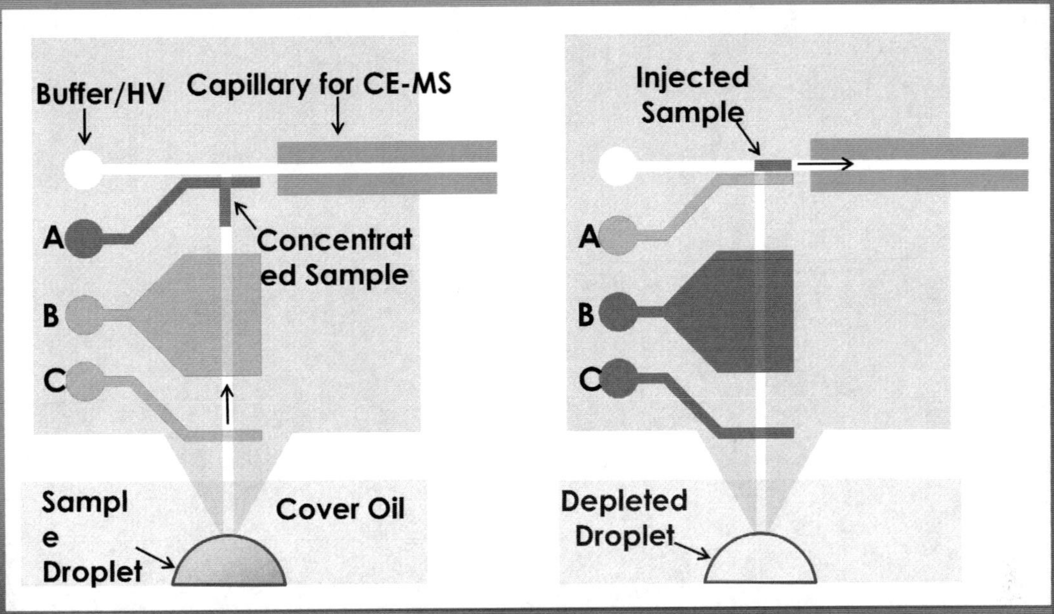

EMSL

Conclusions

- Ultrasensitive NanoESI, PicoESI, MultiESI and chip-based ESI
- Microvalve-based preconcentration and injection for CE and CE-MS
  - No bias
  - Variable sample injection volumes
  - No sample waste
  - Enrichment factors up to 1000
- Potential for rapid, ultrahigh peak capacity LC-CE separations

EMSL

EMSL

Pacific Northwest
NATIONAL LABORATORY

ENERGY

Funding: NIH and US DOE
Yongzheng Cong, Shanta Katipamula, Tao Geng, Cameron Trader, Erin Baker, Danny Orton, Keqi Tang

ENVIRONMENTAL MOLECULAR SCIENCES LABORATORY

431977 Analysis of Amyloid Oligomers Using Microchannel Electrophoresis

Abstract:

Amyloids are involved in a variety of diseases including Alzheimer's Disease and diabetes. A number of technical challenges exist for the quantitative analysis of the different sizes of amyloid proteins which are present during the early stages of aggregation, therefore requiring new technologies to be developed for enhanced protein separations. Microchannel electrophoretic techniques have emerged as powerful tools for the quantitative analysis of proteins. In this study, we demonstrate the importance of coating microchannel walls to reduce protein adsorption as well as the use of field amplified sample stacking. In addition, the effects of sample preparation and oligomers on early aggregation for the model amyloid forming proteins amyloid beta and amylin have been investigated.

437414 Surface Isoelectric Focusing (sIEF) for Qualitative Protein Separation

Abstract:

Surface Isoelectric Focusing (sIEF) for Qualitative Protein Separation

Zhichao Wang, Adrienne R. Minerick

[1]Department of Chemical Engineering, Michigan Technological University, Houghton, MI, 49931, USA

A novel surface isoelectric focusing (sIEF) technique for micro-scale IEF applications is described and utilized to illustrate broad and narrow range ampholyte separations. sIEF is capable of resolving ampholytic molecules such as proteins and peptides from a complex mixture at micrometer scale, which is 100 times smaller than the smallest macrochannel-IEF previously reported. This smaller scale allows sIEF to be seamlessly integrated with protein array libraries. Moreover, the surface-printed property of gel will make it feasible for spot recovery or orthogonal MS, TOF and other secondary analysis.

With sIEF, 40 to 60 μm wide micro-gel line are printed spanning a 300 μm gap between micro-patterned parallel gold electrodes via surface patterning tools (SPTs, Bioforce Nano eNabler^{TM}) on traditional microscope glass slides. Pharmalyte^{TM} carrier ampholyte solution in either broad range (pH 3-10) or narrow range (pH 6.7-7.7) are co-printed with the acrylamide monomer solution and then polymerized in-situ. Mixtures of green fluorescent protein (pI ~6.0) and r-phycoerythrin (pI ~4.4) as well as mixtures of fluorescent peptides pI markers (contains pI 6.8, 7.2 and 7.6) were then printed over the gel surface. Upon electric field applications in the 3 to 9 V range, separation performance tests were quantified. To further optimize the resolving ability of sIEF, dielectric surface coatings were employed to isolate and suppress electrode electrochemical reactions. Also additive chemicals were coprinted with the sIEF gels to reduce protein-protein and/or protein-surface interactions that improve band distortion challenges.

Our results illustrate the development of a 100 pL gel for surface IEF-PAGE with easily adjustable pH gradient ranges. This new sIEF approach has lower applied voltages and smaller sample requirements, relatively quick fabrication process and device reusability, which make it more attractive as a portable, user-friendly platform for qualitative protein separations in a wide variety of protein analysis applications.

428198 mRNA Extraction from Mycobacteria M. Smegmatis Utilizing Ultrahigh Field Intensity Electrolysis

Abstract:

The characterization of gene expression level has been the interest to researchers for a long time, both in specific development stage and physiological condition. Understanding the transcriptome, e.g. the complete set of transcripts in a cell, is essential for interpreting the functional elements of genome. Even though there is not necessarily always a strong correlation between the abundance of mRNA and related proteins, measuring mRNA levels is still informative in understanding how the transcriptional machinery of the cell is affected by external signal or how the cells interact with environment.

Real time PCR, microarrays and RNA sequencing (RNA-seq) have been developed in recent years to quantify the mRNA for understanding physiological events at the whole genome-level level. The purity and integrity of input RNA are critical for the success of these RNA based analysis. Any compromise in RNA quality can lead to variable results. However for some of the species, such as *Mycobacteria*, the evaluation of transcriptome has been hindered by the lysis procedures available. The thick lipid-rich cell wall and slow replication of mycobacteria make chemical disruption of the cells tedious and considerably difficult. Several methods have been reported for the isolation of RNA from mycobacteria, such as the Trizol-based method and the Hot-phenol based method. However, these methods use highly toxic chemicals in several steps and give variable quality of RNA. The typical lysis procedure which is achieved by lysozyme, proteinase K and sodium dodecyl sulfite (SDS) quantity of mRNA typically takes several days. The involved chemicals and physical force may result in the RNA fragmentation which would compromise the accuracy of mRNA evaluation. To minimize fragmentation, the extraction protocol should be rapid with minimal chemical involved. In addition, the efficiency of RNA extraction also needs to be improved to evaluate the absolute amount of RNA in individual cells. The conventional procedure is usually applied to a large number of cells (up to 10^8 cells) which limits its application with limited amount of cells such as bacteria-host interaction or single cell analysis. Hence, it is important to develop new approach for mRNA extraction for mycobacteria, such as *M. Smegmatis*.

Electrical cell lysis shows great advantages of its fast speed and reagent procedure. The electrical lysis is based on electroporation which applies electrical pulses with defined voltages and widths causes the formation of nano-pores on the cell membrane. With increased voltages and duration of electrical pulses, the cells will be lysised and their intracellular materials will be released. To dealing with small number of cells, we conduct electrical lysis in the microfluidic chip. The major challenge of microfluidic electrical lysis is the generation of bubbles due to the water electrolysis. Here we describe a rapid mRNA extraction method utilizing ultra-high field intensity electrolysis. We are able to apply 6,000 V/cm dc field for rapid lysis from about 0.1 million *M. Smegmatis* cells. The mRNA extraction efficiency is about 6 times higher than state-of-the-art bead beating approach.

In our design, we designed a two layer PDMS microfluidic chip for *M. Smegmatis* electrical lysis. The depth of shallow part of the channel (2 μm) is designed to be smaller than microscale silica beads (4.8 μm in the diameter). The suspended microsphere will be stopped by the shallow channel and packed to a microscale matrix that stops the *M. Smegmatis* in the solution. The amount of packed beads is controlled by a pneumatic valve fabricate by multilayer soft lithography. The geometric modification was used to locally amplify the electrical field in the narrow section of the channel, while the rest of the channel is much wider so that the overall voltage required for lysis is relatively low. Subsequent electrical pulses rapid lyse the cells and release intracellular proteins and mRNAs. The decrease of fluorescence from green fluorescent protein (GFP) released by GFP expressing *M. Smegmatis* is observed. The amount of released mRNA is quantified by quantitative reverse transcript polymerase chain reaction (qRT-PCR). In our approach, the capturing and lysis of cells are based on physical methods without involving chemical or biological reagent.

In our approach, the cell lysis procedure is extremely fast without using additional chemicals. The cell lysate can be directly used for qRT-PCR analysis without further treatment. Furthermore, our instrumentation is extremely simple. The electrical pulses was controlled by LabVIEW program and a relay combination. A high voltage power supply was used as a power source. The microfabrication process is simple and low cost. There is no microfabricated electrode used in our design. Finally, our approach can be easily scaled down to analysis small amount of cells and integrated with electrokinetics-based technology.

431344 Electrophysiological Rhythms in Red Blood Cells

Monday, November 9, 2015: 10:30 AM
Ballroom E (Salt Palace Convention Center)

Erin A. Henslee, Mechanical Engineering Sciences, University of Surrey, Guildford, United Kingdom and Fatima H. Labeed, Centre for Biomedical Engineering, Mechanical Engineering Sciences, University of Surrey, Guildford, United Kingdom

Coordination of many of the body's behavioural, metabolic, and physiological processes are controlled

around the 24 hour clock typically controlled through endogenous mechanisms, but can be re-

synchronised (entrainable) by external stimuli. For example, the process of the body overcoming jetlag

via rhythms resetting. Circadian rhythms are observed in most mammalian cells, in vivo and in vitro. From

rhythmic gene expression and cell division, to metabolic oscillations, biological clocks regulate myriad

aspects of cell biology and physiology. Hence, their disruption is strongly associated with disease.

Current models indicate the main drivers of the process are the products of rhythmic expressions of

'clock genes' over a 24 hour cycle, though the specific transcriptional component can differ. O'Neill &

Reddy, 2011 [1] showed following temperature entrainment, circadian rhythms present in human red blood

cells (RBCs). Very little is known about RBC clock mechanisms, or whether functional consequences exist for

erythrocyte biology. To address this, we have established an electrophysiological approach to investigate

the causes and consequences of timekeeping in the RBCs.

We have employed dielectrophoresis through the 3DEP System to simultaneously measure cellular response

at 20 different frequencies to create a DEP "spectrum" of RBCs over a time course [2]. The spectra are then

fit to a single shell dielectric model to obtain electrophysiological properties of the cells.

Following the isolation and entrainment methodologies described in [1], we used DEP to obtain the

electrophysiological properties of RBCs from four male adult donors. Our preliminary results provided DEP

fingerprints of four donors (Figure 1 A-C), from which the extracted membrane conductance Geff

andcytomplasmic conductivity σcyt fluctuated antiphasically with near-24-hour periods. The averages of

Geff and σcyt were fit to cosinor curves peaking at 17.35 h and 5.45 h after the end of the temperature

cycle, respectively.No significant variation was observed in Ceff, confirming the microscopic observation that

cell radius and morphology did not change over the incubation.

The near-antiphasic appearance of Geff and σcyt suggests a rhythmic regulation of

transmembrane ion flux. These results demonstrated, for the first time, a circadian rhythm in the

electrophysiological properties of both red blood cell membrane and cytoplasm. We interpreted the

observed changes in electrophysiological parameters as reflecting alterations in ion channel activity, which

in turn is likely to be mediated by post-translational modifications, in particular phosphorylation. As a first

step towards elucidating the mechanism underlying timekeeping in RBCs, we chose to investigate to what extent it shares key features with the clock mechanism in nucleated cells, As an initial test, we predicted

that RBC rhythms would be sensitive to the generic kinase inhibitor, staurosporine. We observed an increase

in the circadian period of the electrophysiological oscillation.

One of the main issues with detecting circadian desynchrony in physiology is the absence of a non-invasive,

rapid method for detecting systemic temporal state. Recent studies show that forced circadian

desynchrony in humans leads to large-scale perturbation of circadian rhythmicity of blood mRNA's,

suggesting that blood is a good reflection of the wider internal desynchrony. Measurement of circadian

rhythms in animals routinely requires the sacrifice of multiple animals per time-point, meaning that tens of

animals are used in each experiment. DEP has the potential to provide an accurate and reproducible readout

of circadian status using very small blood volumes, and significantly reduce the numbers of animals required

for an experiment. Our preliminary data, using 10 μL tail-knick whole blood samples from voles,

demonstrates, for the first time, an electrophysiological correlate of the ultradian feeding-fasting cycle in

blood (Figure 1 D&E).

In the absence of nuclei or other organelles in RBCs, this work potentially indicates membrane

mechanisms involving differential ion channel activities and renders DEP a new technology to enhance the

understanding of chronobiology. Future work will investigate the underlying mechanisms behind these changes and the effects of experimental conditions.

1.)O'NEILL, J. S. & REDDY, A. B. 2011. John S. Circadian clocks in human red blood cells. *Nature,* 469,

498–503

2.) HOETTGES, K. F., HÜBNER, Y., BROCHE, L. M., OGIN, S. L., KASS, G. E. N. & HUGHES, M. P. 2008.

Dielectrophoresis-activated multiwell plate for label-free high-throughput drug assessment.

Analytical Chemistry, 80**,** 2063-2068.

3.) IRIMAJIRI, A., HANAI, T. & INOUYE, A. 1979. Dielectric theory of multi-stratified shell-model with

its application to a lymphoma cell. *Journal of Theoretical Biology,* 78**,** 251-269.

4.) FRY, C.H., SALVAGE, S.C., MANAZZA, A., DUPONT, E., LABEED, F. H., HUGHES, M. P. & JABR, R.I.

2012. Cytoplasm Resistivity of Mammalian Atrial Myocardium Determined by Dielectrophoresis and

Impedance Methods. *Biophysical Journal,* 103(11), 2287–2294.

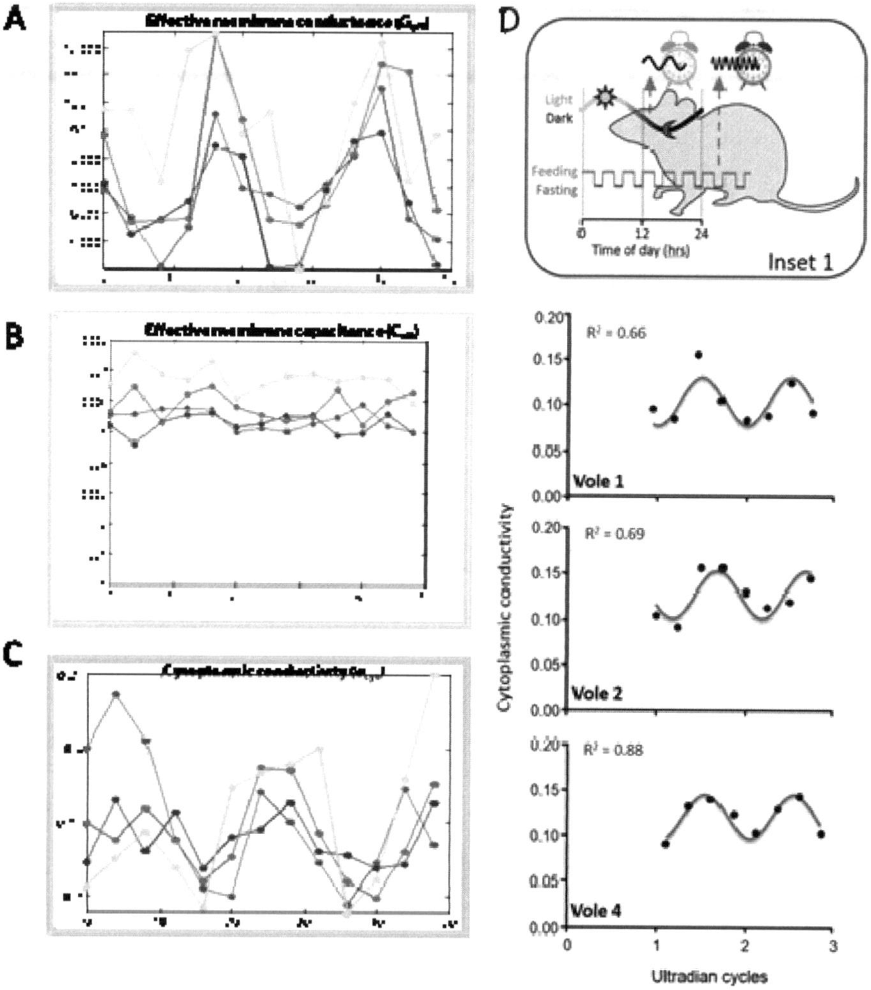

Figure SEQ Figure * ARABIC 1: Electrophysiological results for isolated human RBCs. Circadian variation in A.) Membrane conductance, B.) Membrane Capacitance, and C.) Cytoplasmic Conductivity. D.) Ultradian variation in vole whole blood samples of cytoplasmic conductivity.

437259 On the Influence of Outer Membrane Cytochromes on Cell Polarizability

Abstract:

On the Influence of Outer Membrane Cytochromes on Cell Polarizability

Qianru Wang[1], Pei Zhang[1], A-Andrew Jones III[1], Liwei Lin[2] and Cullen R. Buie[1]

[1]Department of Mechanical Engineering, Massachusetts Institute of Technology

[2]Department of Mechanical Engineering, University of California, Berkeley

This work elucidates for the first time the strong correlation between extracellular electron transfer (EET) facilitated by outer membrane cytochromes and cell envelope polarizability. EET is a crucial cell phenotype for applications including environmental remediation, microbial fuel cells (MFC), and microbial electrosynthesis. Expression of outer membrane c-type cytochromes plays an important role in bacteria respiration and EET. Although various redox active anode-reducing cytochromes critical for EET in *Geobacter sulfurreducens* (*G. sulfurreducens*) have been investigated[1-3], and microfluidic systems have been explored to measure cellular electrical properties[4,5], the correlation between the cellular physiological properties and their electrical phenotype has not been revealed. In this work, analysis of wild type *G. sulfurreducens* and cytochrome deletion mutants show that expression of outer membrane cytochromes, which are responsible for EET, influence cell envelope polarizability as measured in low cost microfluidic systems employing dielectrophoresis (DEP). The strong correlation between EET and cell surface polarizability can be exploited to identify novel electrochemically active bacteria, with broad implications for biotechnology, bioremediation, and biomedical applications.

We find that the expression of periplasmic and outer membrane bound c-type cytochromes has a significant effect on *G. sulfurreducens* surface polarizability. Generally speaking, strains expressing larger amounts of c-type cytochromes display higher polarizability. However, depending on the specific function of the c-type cytochrome, the various genes explored influence cell envelope polarizability differently. C-type cytochromes important for oxidation/reduction reactions, but not actively involved in electron transfer, appear not to influence cell envelope polarizability. Furthermore, simultaneous deletion of multiple outer membrane bound c-type cytochrome genes doesn't result in linear reduction of cell polarizability due to the strain's ability to adapt increased expression of other cytochrome genes. In conclusion, for the first time the electrochemical phenotype of cells is definitively correlated with cell polarizability. This work demonstrates the power of microfluidic techniques such as DEP for microbial biotechnology. Future work in this area could extend to the development of novel microfluidic tools employing DEP to identify or evolve microbes for bioelectrochemical systems and other applications dependent upon EET.

Word: 333

REFERENCES:

1. Bond, D. R.; Lovley, D. R., *Appl. Environ. Microbiol.* **2003,** 69, (3), 1548-1555.

2. Holmes, D. E.; Chaudhuri, S. K.; Nevin, K. P.; Mehta, T.; Methe, B. A.; Liu, A.; Ward, J. E.; Woodard, T. L.; Webster, J.; Lovley, D. R., *Environmental Microbiology* **2006,** 8, (10), 1805-1815.

3. Inoue, K.; Qian, X. L.; Morgado, L.; Kim, B. C.; Mester, T.; Izallalen, M.; Salgueiro, C. A.; Lovley, D. R., *Applied and Environmental Microbiology* **2010,** 76, (12), 3999-4007.

4. Hawkins, B. G.; Huang, C.; Arasanipalai, S.; Kirby, B. J., *Anal. Chem.* **2011,** 83, (9), 3507-3515.

5. Zhao, Y.; Chen, D. Y.; Luo, Y. N.; Li, H.; Deng, B.; Huang, S. B.; Chiu, T. K.; Wu, M. H.; Long, R.; Hu, H.; Zhao, X. T.; Yue, W. T.; Wang, J. B.; Chen, J., *Lab Chip* **2013,** 13, (12), 2272-2277.

436826 Monitoring Microbial Hunger Games: Dielectrophoretic Monitoring of Inter-Strain C.Difficile Antagonisms

Abstract:

Clostridium difficile (*C.difficile*) infection (CDI) is a global toxin-mediated intestinal disease that is commonly attributed to exposure to pathogenic *C.difficile* strains following the elimination of healthy microflora in the gut by broad-spectra antibiotics. An emerging body of scientific and clinical studies has demonstrated that CDI rates can be controlled through promoting antagonistic inter-strain interactions on pathogenic C.difficile strains by pre-colonization with non-toxigenic *C.difficile* and other probiotic microorganisms, due to ensuing competition for the same nutrient and colonization niche. The emerging interest in restoration of a healthy microbiota to treat CDI highlights the need to understand and optimize the complex and dynamic interaction between the commensal polymicrobial environment and pathogenic *C.difficile* strains. Current methods to monitor the alterations in C.difficile colonization following culture within environments including non-toxigenic *C.difficile* and other probiotic microorganisms utilize cumbersome adhesion assays on human epithelial cell lines (HCT8). These are time consuming, thereby limiting the permutations of interactions that can be studied and do not provide direct information on the causes underlying the alterations in colonization. S-layer glycoproteins are part of the cell wall envelope in gram positive and gram negative bacteria, and are integral to surface recognition, gut colonization, host-pathogen adhesion and virulence. Hence, alterations in colonization properties of pathogenic *C.difficile* strains due to dynamical interactions with non-toxigenic *C.difficile* and other probiotic microorganisms are likely to alter the S-layer on the cell wall envelope. We demonstrate here that these alterations in the S-layer on the cell wall envelope of pathogenic *C.difficile* strains due to interactions with non-toxigenic *C.difficile* cause systematic alterations in call wall capacitance [1], thereby enabling their facile monitoring within just as few minutes, based on the characteristic shifts in their dielectrophoretic frequency spectra. We compare the influence of static versus dynamic co-culture methods on alterations in *C.difficile* colonization and benchmark this dielectrophoretic method versus conventional adhesion assays. In this manner, we envision the application of this microfluidic platform towards optimization of polymicrobial therapies to arrest CDI.

[1] Y.-H. Su, C. Warren, R.L. Guerrant, N.S. Swami; "Dielectrophoretic monitoring and inter-strain separation of intact *Clostridium difficile* based on their S(surface)-layers", *Anal Chem* (2014), 86, 10855–10863.

417364 Dielectrophoretic Separation of Babesia-Infected Erythrocytes

Abstract:

Babesia species are obligate intraerythrocytic tick-borne protozoan parasites that are the etiologic agents of babesiosis, a potentially life-threatening, malaria-like illness in humans. Babesia-infected people have been known to suffer from complications including liver problems, severe hemolytic anaemia and kidney failure. As reported to the Food and Drug Administration, human Babesiosis accounts for almost all of the 38% mortality cases observed in transfusion recipients. As of now, no tests have been licensed yet for screening blood donors for Babesiosis. Current diagnostic tools for Babesiosis including Enzyme-Linked ImmunoSorbent Assay (ELISA), Fluorescence in Situ Hybridization (FISH), Polymerase Chain Reaction (PCR) are expensive and burdened with multifarious shortcomings. We present, here, a low-cost, high-specificity, quick, and easy-to-use electrokinetic diagnostic tool for separating *Babesia*-infected cells. In this work, we exposed a mixture of Babesia-infected (varying parasitemia) and healthy red blood cells to non-uniform electric fields in a fabricated microfluidic platform to manipulate and isolate the *Babesia*-infected cells within a minute. At DC voltage configurations of 10V, 0/6V in the inlet and two outlet channels respectively, the diseased cells were seen to flow in a direction different from the healthy red blood cells'. This result shows that our microdevice, when fully developed into a point-of-care, lab-on-a-chip diagnostic device, could be used to screen donors' blood for Babesiosis at donation centers, thus preventing contamination of the blood bank. The project is ongoing and future work will focus on differentiating the various other species of Babesia and their co-infecting agents.

431424 Optimizing Neural Stem Cell Sorting with Dielectrophoresis

Abstract:

Human neural stem cells have enormous potential as therapeutics for central nervous system injuries and diseases. However, cultures of these cells are heterogenous, containing a mix of stem cells and progenitors linked to specific fates. Distinct progenitors in the neural lineage are detected in dielectrophoresis (DEP) without the use of labels, creating a rapid and convenient method for analyzing and separating these cells.

We demonstrated previously that astrocyte and neuron progenitors could be isolated from mouse neural stem and progenitor cells (NSPCs) at specific frequencies in DEP. Furthermore, the specific membrane capacitance values of both mouse and human cells reflect their fate potential and distinguish astrocyte and neuron progenitors. In our previous DEP-based sorts, we utilized a set frequency range to isolate a specific progenitor population. However, we found that the cells' responses to the frequencies were not always the same, potentially due to minor variations in the batches of cells, devices used for sorting, or media conductivity across different sorts.

To address this issue, we compared predicting the appropriate sorting frequency by two approaches - generating a pre-sorting analysis of cell trapping at escalating frequencies in the sorting device (mini trapping curve) or analyzing the cells in a 3DEP Reader. Data obtained from the 3DEP Reader analysis provided a good prediction of the mini trapping curve data and rapidly identified frequencies for sorting. Human NSPCs sorted by frequencies identified in the 3DEP analysis were enriched as confirmed by post-sorting DEP analysis indicating shifts in the DEP frequency response and membrane capacitance of sorted cells compared to controls. Progenitor cells sorted using this approach were analyzed for fate potential by assessing differentiation into neurons or astrocytes. Further analysis focused on quantitative assessment of N-glycosylation in sorted progenitors since we previously found clear links among cell surface glycosylation, membrane capacitance, and cell fate in the neural lineage. Efficient sorting of human neural stem cells into distinct pools of progenitors enables further study and optimization of these cells to treat human injuries and disease.

434622 Identification of Various Stages of Induced Apoptosis and Treatment Sensitivity in Non-Small Cell Lung Carcinoma Using Dielectrophoresis

Abstract:

It is important to understand various biomolecular events that contribute towards classification of programmed cell death (apoptosis) and how tumors evade apoptotic death. Small cell lung carcinoma (SCLC) and non-small cell lung carcinoma (NSCLC) are two major categories of lung cancer that differ in their sensitivity to undergo apoptosis. With new drugs being formulated to control abnormal cell proliferation (cancer) caused by mutations, effective drug-induced apoptosis will yield the relationship between cancer genetics and treatment sensitivity. The objective of this study was to build non-invasive techniques (without altering inherent cell properties) to isolate cells that are undergoing varying stages of apoptosis for a high-throughput screening application. We have used dielectrophoresis to determine and isolate various stages of apoptotic cells (early, mid and late) for two NSCLCs adenocarcinoma cell lines (HCC1833 and H1755). Our studies have shown significant differences in apoptotic cells by chromatin condensation, formation of apoptotic bodies and exposure of phosphatidylserine (PS) on the extracellular surface when the cells where exposed to a potent Bcl-2 family inhibitor drug (ABT-263). Time lapse dielectrophoretic studies were performed over a period of 48 hours upon exposure of varying concentrations of ABT-263 ranging between 50 nM to 500 nM. As a result of physical and biochemical changes, inherent dielectric properties of cells undergoing varying stages of apoptosis showed amplified changes in their cytoplasmic and membrane capacitance. In addition, zeta potential of these fixed isolated cells were measured to obtain direct correlation to biomoelcular events. Western blot was used to examine the changes in the expression levels of Bax, Bcl-2, caspase-3, Hsp-90. As a control, the obtained results were compared against standard apoptotic assays for respective stages; JC-1, a mitochondrial membrane potential dye for marking early stage apoptosis, Annexin-V for the detection of PS as a mid-stage marker and defragmentation of DNA, verified by immunofluorescence staining using fluorescein-deoxyuridnie triphosphate (FITC-dUTP) and flow cytometry. The use of non-invasive dielectrophoresis is to provide a sophisticated method to characterize and isolate cells for potential downstream analysis. Effective separation of carcinoma cells at different apoptotic stages should enable a more rational approach to anticancer drug design and therapy.

432008 Robust Dielectrophoretic Cell Aggregation in Biocompatible Hydrogels

Abstract:

The ability to build a three-dimensional (3D) cellular construct is a field of increasing interest. Current cell-based in vitro studies are typically two-dimensional (2D) monolayer models that have been shown to be unreliable and non-predictive in a clinical setting as they do not represent human tissue models [1-2]. In vitro 3D model development has been brought into focus in order to develop a system that can resemble the natural living tissue, allowing cells to assume their natural shape and interaction without the need for animal models. In comparison to 2D models, 3D cell culture allows cells to assume their natural shape, and allows cell-cell interactions which can affect disease progression and drug responses in cells [2]. Disease progression and drug efficacy have been shown to vary between 2D versus 3D constructs [3]. There is also debate whether the mechanism underpinning this increase in drug resistance is due to changes in the behaviour of cells due to cell-to-cell contact, or simply that the inner cells in a 3D construct are shielded from the drug due to the outer cells preventing diffusion of the drug across the whole cell mass.

Proposed systems for 3D spheroids (most flexible and well characterized in vitro model) formation such as spinners, shakers, rotaries, as well as hanging drop method are simple to use and designed for mass fabrication. These however, have difficulties maintaining a uniform size and geometry in micro-scale spheroids which are crucial when investigating avascular effects in cell constructs [4-6]. The use of dielectrophoresis (DEP) to construct hemispherical cell clusters in polymer hydrogels have shown promise [6-8] however clinical application of this technology, such as drug efficacy has yet to be shown. A common method of representing this drug efficacy is the half maximal inhibitory concentration (IC_{50}) which measures the amount or dose of drug that is able to inhibit a specific biological process (such as cell division leading to cell death). MTT is the current method of obtaining the IC_{50} of a drug on a cell line grown in traditional monolayer culture. Since this method relies on cell suspensions, other methods must be used on 3D aggregates.

In this work we present an on-chip DEP device as a robust, high through-put, reproducible technique of cell aggregation and assessment of drug effectiveness. Using "dot" electrodes, we successfully aggregated and maintained 3D culture for several cell lines including yeast, k562 human leukaemia cells, cardiomyoctyes, and HeLa's. Further, the effect of Amphotericin B on patterned yeast cells and doxorubicin on the survival and proliferation of patterned and encapsulated cancer cell aggregates were observed. Comparisons of the 2D versus 3D IC_{50} of these agents were investigated. In demonstrating DEP as a robust technique for cell aggregation, alternative hydrogels such as collagen and PuraMatrix™ were also investigated to provide a more biocompatible method of cell aggregation and to allow further study of aggregates through dissociation. The DEP dot electrodes have demonstrated the potential for rapid bench top cell aggregate formation and have shown promise in clinical application

References:

1. Elliot.N.T and Yuan.F, A Review of Three-Dimensional In Vitro Tissue Models for Drug Discovery and Transport Studies. Journal of Pharmaceutical Sciences, 2010. 100(1): p. 59-74.

2. Khetan.S and Burdick.J.A, Patterning hydrogels in three dimensions towards controlling cellular interactions. Soft Matter, 2011. 7: p. 830-838.

3. Petersen.O.W, et al., Interaction with basement membrane serves to rapidly distinguish growth and differentiation pattern of normal and malignant human breast epithelial cells. Proceedings of the National Academy of Sciences of the United States of America, 1992. 89(19): p. 9064-9068.

4. Hye-Jin Jin, Young-Ho Cho, Jin-Mo Gu, Jhingook Kim and Yong-Soo Oh, A multicellular spheroid formation and extraction chip using removable cell trapping barriers. Lab chip, 2010, 11,115-119

5. Heike Hardelauf, Jean-Philippe Frimat, Joanna D. Stewart, Wiebke Schormann, Ya-Yu Chiang, Peter Lampen, Joachim Franzke, Jan G. Hengstler Cristina Cadenas, Leoni A. Kunz-Schughart and Jonathan West, Microarrays for the scalable production of metabolically relevant tumour spheroids: a tool for modulating chemosensitivity traits, Lab Chip, 2011, 11, 419–428

6. Rula G. Abdallat, Aziela S. Ahmad Tajuddin, David H. Gould, Michael P. Hughes, Henry O. Fatoyinbo, Fatima H. Labeed, 2013. Process development for cell aggregate arrays encapsulated in a synthetic hydrogel using negative dielectrophoresis. Electrophoresis volume 34 pp 1059-1067.

7. Albrecht, D. R., Underhill, G. H., Mendelson, A. & Bhatia, S. N., 2007. Multiphase electropatterning of cells and biomaterials. The Royal Society of Chemistry, Volume 7, pp. 702-709.

8. Agarwal S, Sebastian A, Forrester LM, Markx G. Formation of embryoid bodies using dielectrophoresis. Biomicrofluidics. 2012 Vol. 6, No. 2, 024101, 03.04.2012.

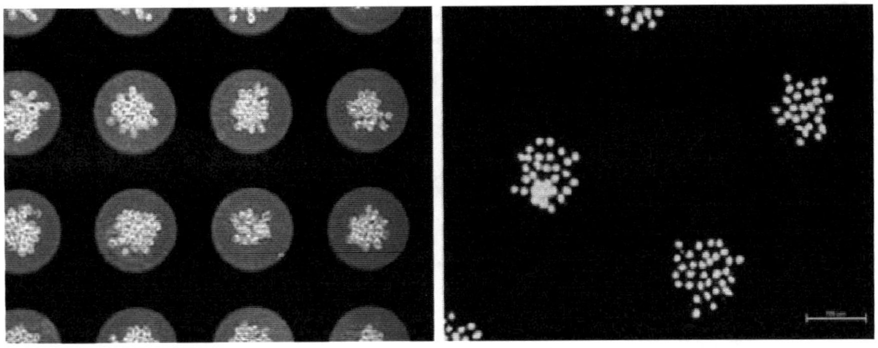

Figure SEQ Figure * ARABIC 1: HeLa cells aggregated on dot electrodes at time 0, and viability tested 48 hr post DEP exposure

434977 Measuring Electrical Properties of Cancer Cells Via Electrorotation

Monday, November 9, 2015: 2:30 PM
Ballroom E (Salt Palace Convention Center)

Timothy Lannin[1], Chao Huang[2], Fredrik Thege[2], Conor Gruber[3] and Brian J. Kirby[1,2,4], (1)Sibley School of Mechanical & Aerospace Engineering, Cornell University, Ithaca, NY, (2)Biomedical Engineering, Cornell University, Ithaca, NY, (3)Cornell University, Ithaca, NY, (4)Division of Hematology and Medical Oncology, Weill Cornell Medical College, New York, NY

Measuring Electrical Properties of Cancer Cells via Electrorotation

Circulating tumor cells are cancer cells that can be extracted from patient blood, and their capture provides a window through which we can examine the progression of cancer in real time with minimal invasiveness. As these cells circulate, they are exposed to a variety of physical and chemical stimuli that we hypothesize change the structure and composition of the membrane, nucleus, and cytoplasm. For example, the morphology of the cell membrane is sensitive to the deprivation of growth factors or the acquired resistance to Gemcitabine chemotherapy. Because these cellular changes are dynamic, it is critical to characterize this process in order to effectively capture and analyze the cells.

We have conducted experiments to understand how stimuli that cancer cells experience in circulation affect their electrical phenotype in order to 1) gain insights into cancer cell membrane physiology, which will help reveal connections between primary tumors, circulating cells, and metastases, and 2) inform better cell capture with electrokinetic techniques, such as dielectrophoresis (DEP), which will enable new sets of genetic and other analyses that require higher purity than the current state of the art in cell capture provides.

Although DEP is suitable for applying forces to cells for separations, electrorotation, a technique related to DEP in which torques are applied to cells in response to rotating electric fields, is more suitable for experimental characterization of cells. An electrorotation spectrum gives complete information necessary to infer a DEP spectrum and extract electrical properties such as the cytoplasmic conductivity and membrane specific capacitance.

We present data on changes in pancreatic cancer cell lines' electrorotation spectra after serum starvation, a stress known to alter cell membrane morphology and cytoplasmic contents. (See attached figure for electrorotation spectra of serum-starved and serum-supplemented BxPC3 pancreatic cancer cells). We also present electrorotation spectra of a Gemcitabine-resistant pancreatic cancer cell subclone, which we compare to their Gemcitabine-naïve counterparts.

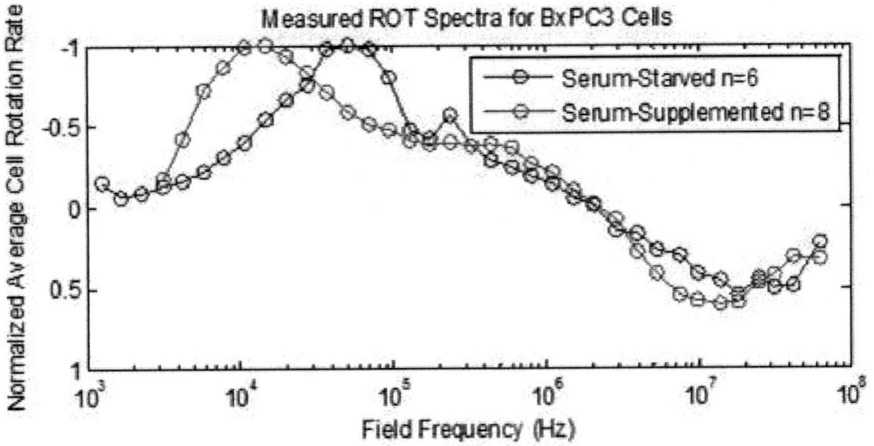

Measured ROT Spectra for BxPC3 Cells
Normalized Average Cell Rotation Rate
Field Frequency (Hz)
Serum-Starved n=6
Serum-Supplemented n=8
-1
-0.5
0
0.5
1
10^3
10^4
10^5
10^6
10^7
10^8

418583 One Step Microfluidic Immunomagnetic Separation of Tumor Initiating Cells Based on Multiple Markers

Abstract:

There has been increasing proposition and observation that cancer growth is driven and sustained by tumor initiating cells (TICs, also known as cancer stem cell (CSC)), which are capable of self-renewal and aberrant differentiation. TICs usually express abnormal high or low level of proteins, but their detection and isolation present unique challenges due to their extremely low frequency (<1%) in most human tumors. Microfluidics has been used for rare cell separation because of their minimal sample requirements. However, current methods are limited in sorting of TICs based on single marker in one step, which is not sufficient for high purity selection; or multiple markers in multiple steps, which increases the risk of cell loss. Here, we propose a new strategy that allows isolation of TICs in one step using two beads. Breast cancer TICs are identified as expression of $CD44^{+}/CD24^{-}$. In our method, $CD24^{+}$ cells are excluded by firstly occupying their surface with CD24 antibody linked nonmagnetic beads, so that CD44 antibody linked magnetic beads are only able to bind to $CD24^{-}$ cells. Finally, $CD44^{+}/CD24^{-}$ cells are enriched after one step microfluidic immunomagnetic separation. We tested our technology with SUM 149 cell line. Cells enriched with two beads method generate 2.4 times tumorspheres compared to unsorted cells, and they showed 30% more efficiency in tumorsphere formation compared to one beads method. We also test unsorted and sorted cells with flow cytometry. CD44 antibody coated magnetic beads select cells with high expression of CD44, while cells sorted with our two beads method shift to both $CD44^{+}$ direction and $CD24^{-}$ direction. We envision that our approach will be useful for highly specific selection of TICs with two markers of opposite preferences. Our method allows the isolation of TICs with high purity from small samples, thus will provide a useful tool for cancer diagnostic and therapeutic researches on precious samples (e.g. primary cells from animals and patients).

437854 Deterministic Absolute Negative Mobility for Sub-Micrometer Bioparticle Separation

Abstract:

Deterministic Absolute Negative Mobility for sub-micrometer Bioparticle Separation

Jinghui Luo[1], Katherine Muratore[3], Edgar Arriaga[2,3] & Alexandra Ros[1]

[1]Department of Chemistry and Biochemistry, Arizona State University, Tempe, Arizona, USA.

[2]Department of Chemistry, University of Minnesota, Minneapolis, Minnesota, USA.

[3]Department of Biochemistry, Molecular Biology and Biophysics, University of Minnesota, Minneapolis, Minnesota, USA. (Alexandra.Ros@asu.edu)

Efficient particle separation in the micron- and submicron range is required for the various analytical challenges related to nanotechnology and microbiology. Here, we demonstrate a non-intuitive, yet efficient migration mechanism for µm- and sub-µm colloidal particles and organelles, taking advantage of particle transport in a nonlinear post array in a microfluidic device under the periodic action of electrokinetic and dielectrophoretic forces. We reveal regimes in which deterministic particle migration opposite to the average applied force occurs for a larger particle – a typical signature of deterministic absolute negative mobility (dANM) -whereas normal response is obtained for smaller particles. The coexistence of dANM and normal migration was characterized and optimized in numerical modeling and subsequently implemented in a microfluidic device demonstrating at least two orders of magnitude higher migration speeds as compared to previous ANM systems. We could also induce dANM for mouse liver mitochondria and envision that the efficient size selectivity can be advantageously employed in organelle sub-population studies, for nanotechnological applications, or in fractionating protein nanocrystals.

416792 A Bias-Free Method for Concentrating Proteins

Abstract:

Human samples for protein profiling often are pretreated to reduce their dynamic range. Common methods include immunodepletion (such as Agilent's MARS column) or concentration-leveling (such as Bio-Rad's ProteoMiner). In either case, the proteins recovered from their pretreatment benefit substantially from concentration.

We first review existing methods to perform this concentration. These include precipitation with trichloroacetic acid, total precipitation with large acetone volumes, use of membrane concentrators, and many others. All of these have inherent biases, often to very dilute species and those of low molecular weight, such as peptides.

Presented here is a bias-free electrophoretic method to perform such sample concentrations. Both theory and experiment are included; the emphasis is on physical principles rather than mathematical details.

CLEMSON UNIVERSITY

Enrichment of small cell populations from large sample volumes using 3D Carbon-electrode Dielectrophoresis

Monsur Islam[1], Maria Fernanda Larraga-Martinez[2]
Rucha Natu[1], Rodrigo Martinez-Duarte[1]

[1]Mechanical Engineering Department, Clemson University, Clemson, SC, USA
[2]Biomedical Engineering Department, University of Iowa, Iowa City, IA, USA

Problem Statement

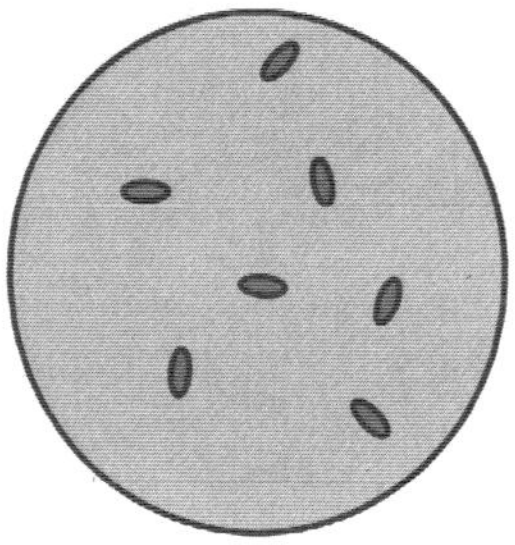

Very low concentration of pathogen in blood

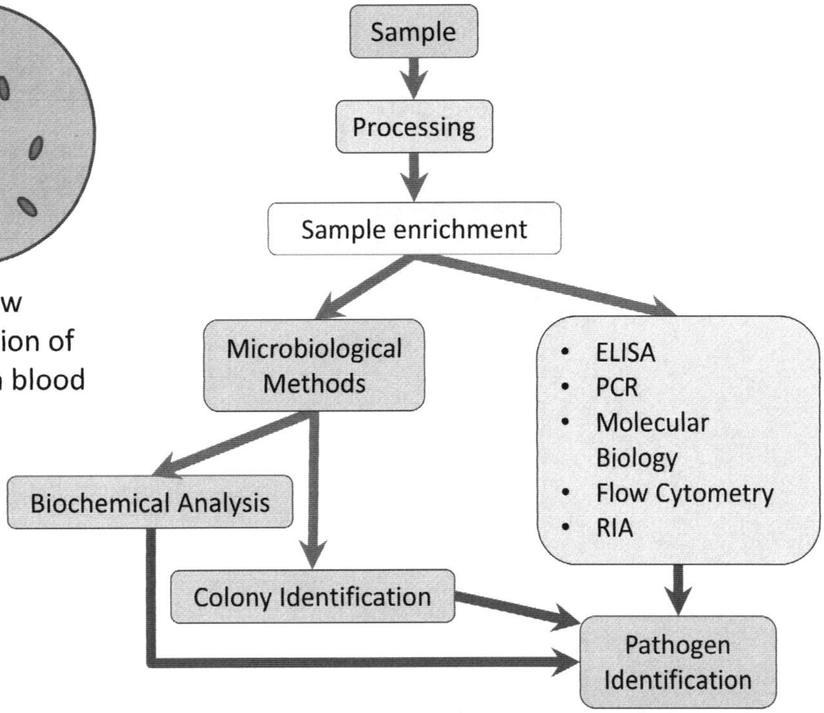

Ref: Singh et al., *Sensors* (2013), 13(2), 1763-1786

Sample Preparation

The vision

Sample in

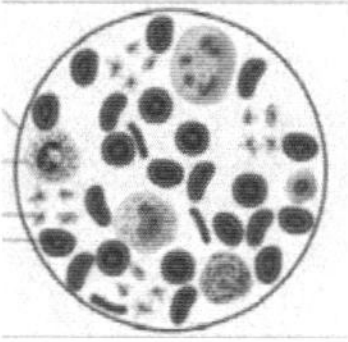

30 ml of whole blood

Result

Hundreds of μl

10 μl

Separation of particles resembling bacteria	Viable Bacteria concentration and purification	Pathogen Identification
Centrifugation Using Bingham Plastics	3D carbon-electrode Dielectrophoresis	Biosensors

www.multiscalemanufacturing.net 3

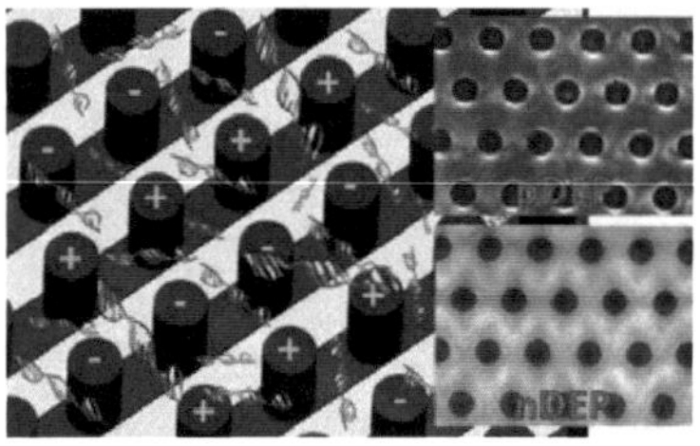

DNA trapping

Arizona State University

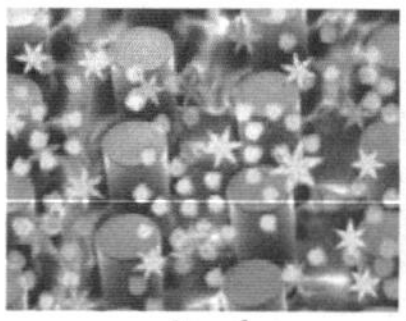

Lysis

École Polytechnique Fédérale de Lausanne

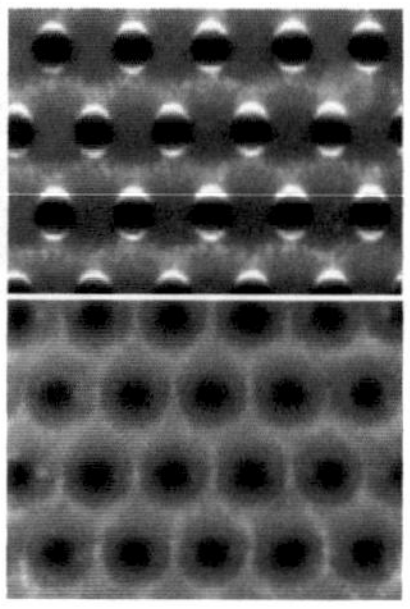

Smeg
for TB research

Sample Preparation Applications

Universitat de Barcelona

Removal of contaminants to facilitate Molecular Diagnostics

cytognos

Sample Prep for Flow cytometry

www.multiscalemanufacturing.net 4

Parameters of interest

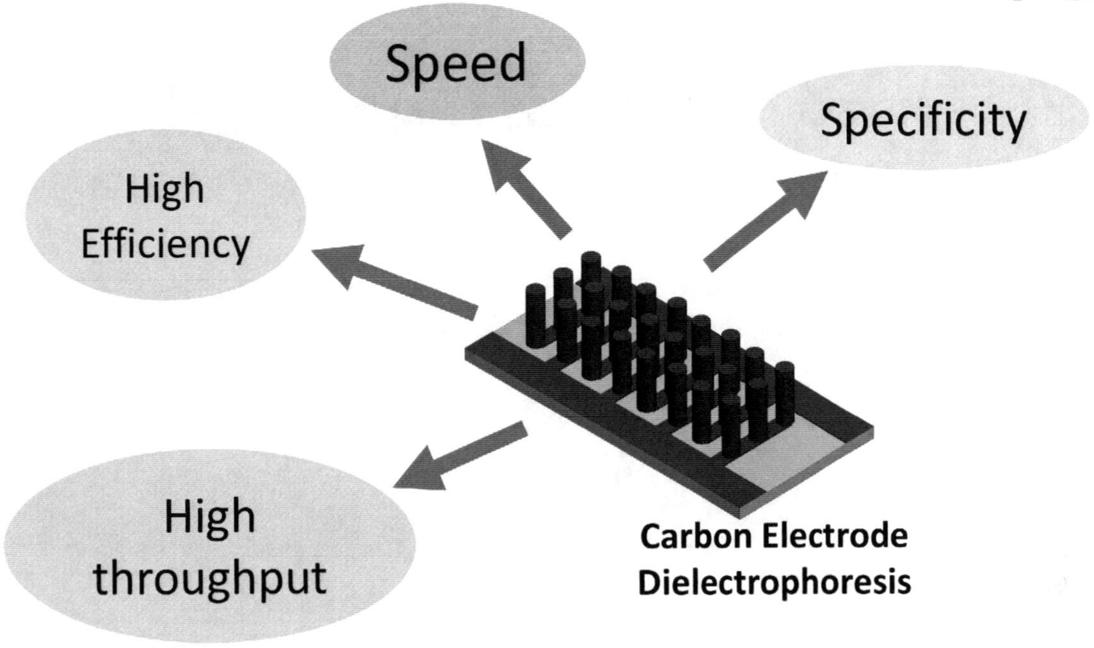

Dielectrophoresis(DEP)

DEP - Interaction of dielectric particle with non-uniform electric field

$$F_{DEP} = 2\pi r^3 \varepsilon_m \, Re(f_{CM}) \, \nabla E_{rms}^2$$

$$Re(f_{CM}) = \frac{\varepsilon_p^* - \varepsilon_m^*}{\varepsilon_p^* + 2\varepsilon_m^*} \longrightarrow \varepsilon^* = \varepsilon - j\left(\frac{\sigma}{\omega}\right)$$

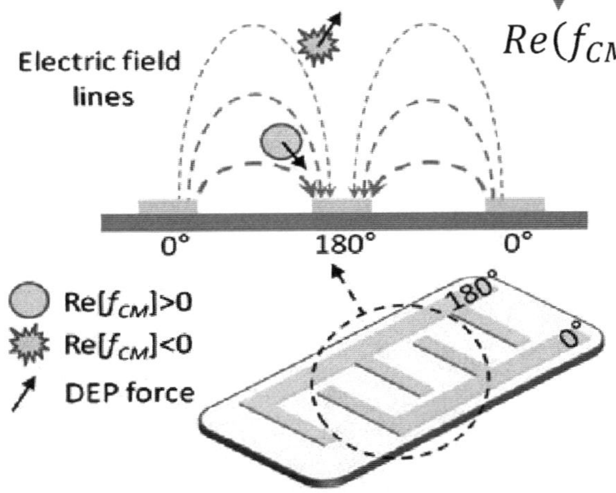

$Re[f_{CM}] > 0 \longrightarrow$ Positive DEP

$$F_{DEP} > F_{Drag}$$

$Re[f_{CM}] < 0 \longrightarrow$ Negative DEP

Ref: Khoshmanesh et al., Biosensors and Bioelectronics, 26 (2011) 1800–1814

Fabrication of DEP Device

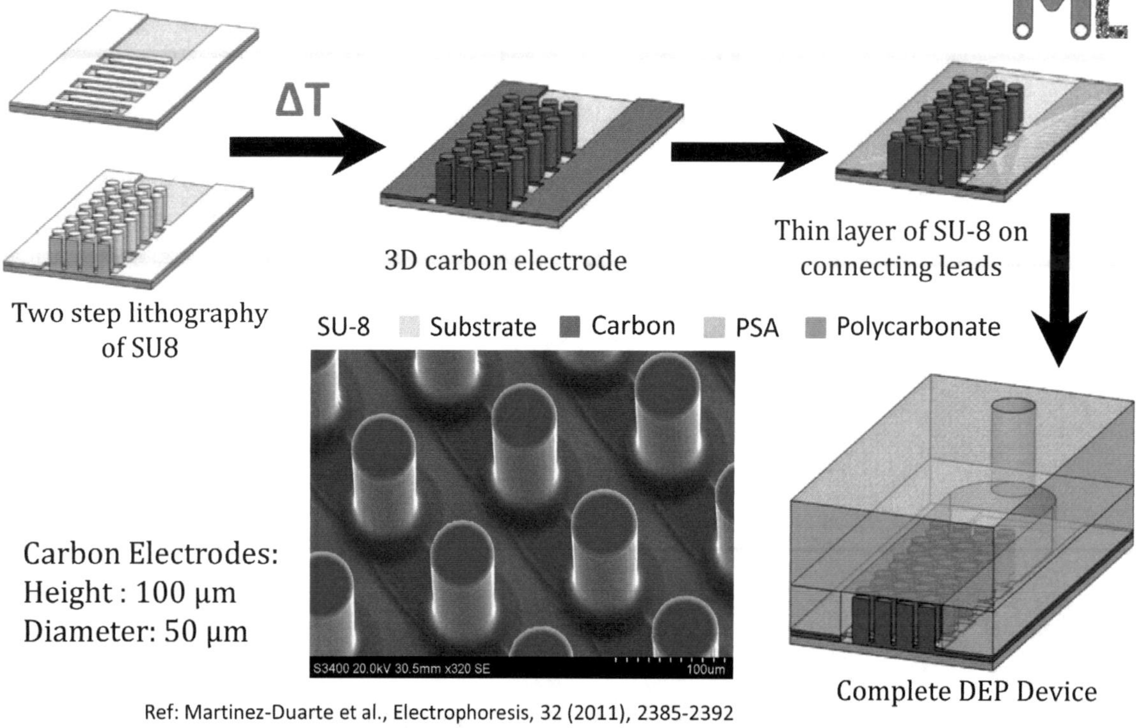

Ref: Martinez-Duarte et al., Electrophoresis, 32 (2011), 2385-2392

Experimental Procedure

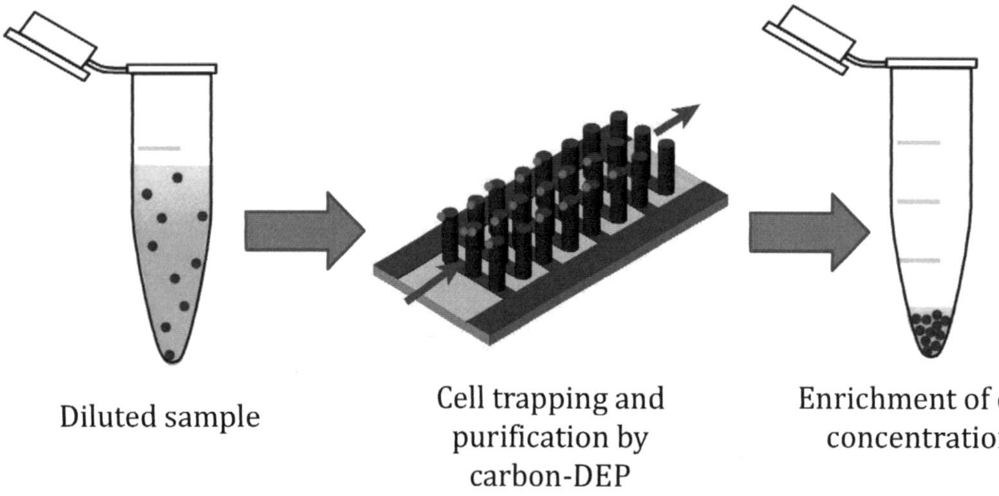

DEP Protocol

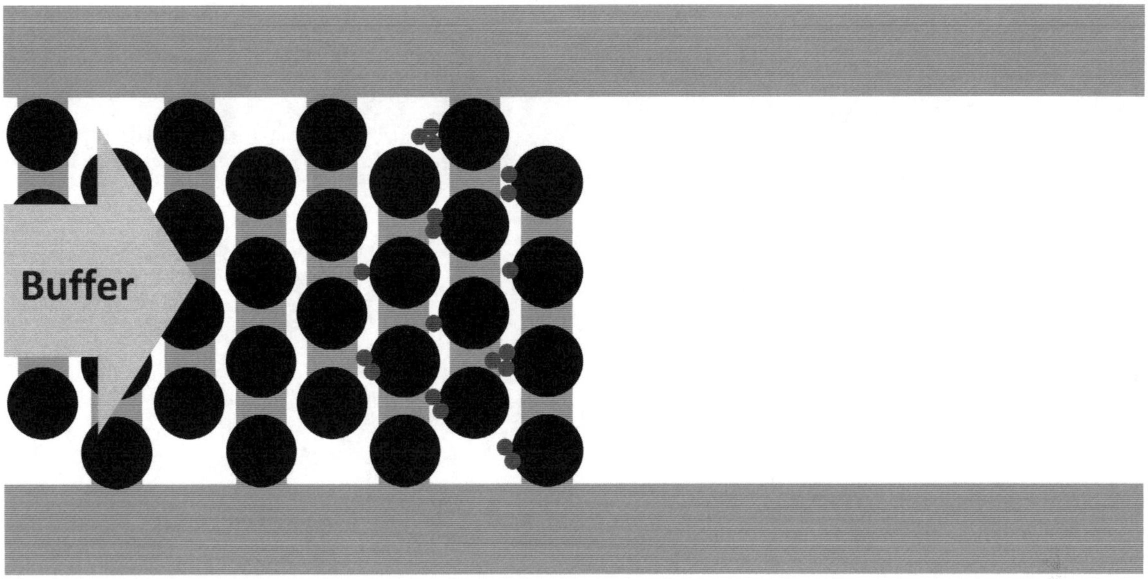

Signal On

DEP Protocol

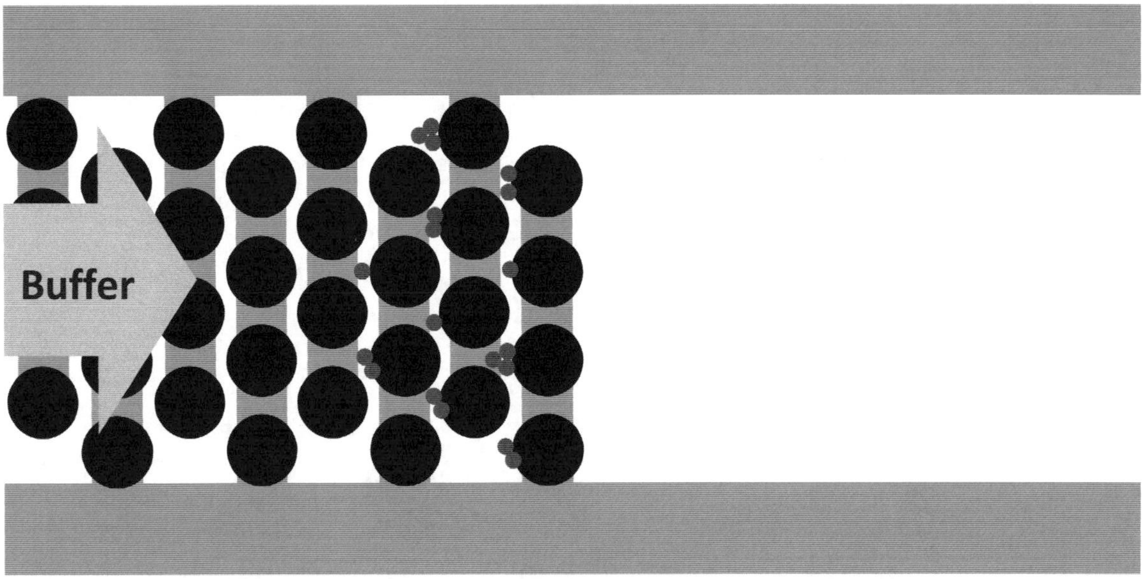

Signal Off

Results with different cell concentration

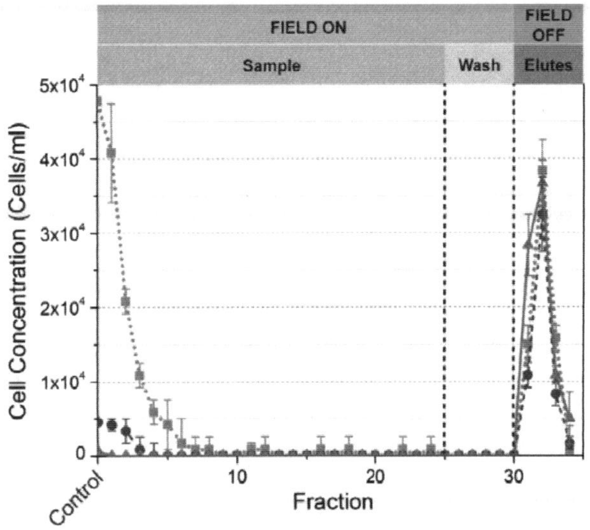

Minimum number of Cells= 200
Initial cell concentration and sample volume:

- 10^4 cells/ml and 500 µl
- 10^3 cells/ml and 500 µl
- 10^2 cells/ml and 4 ml

- Flow rate = 10 µl/min
- Vpp = 20 V
- Frequency = 100 kHz

- Cell saturation in the chip ~ 10^4 cells/ml.
- Cell enrichment was achieved up to ~ 3.5 X 10^4 cells/ml

Results with different flow rate

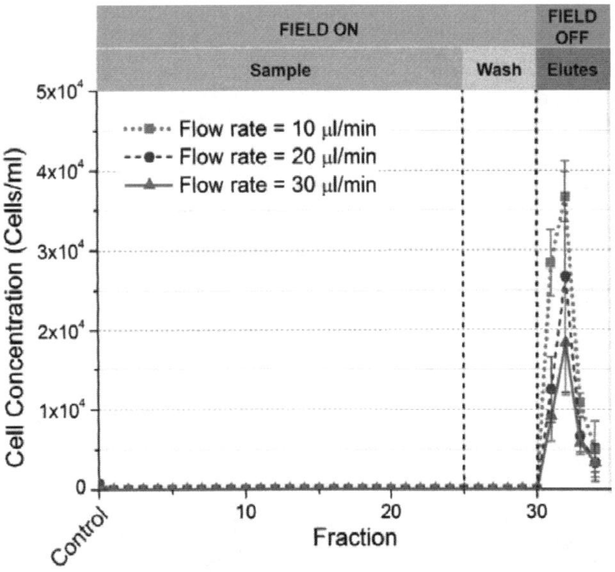

- Initial cell Concentration= 10^2 cells/ml
- Vpp = 20 V
- Frequency = 100 kHz
- Sample volume = 4 ml

Flow rate (µl/min)	Duration of the sample processing
10	6 hours 50 minutes
20	3 hours 38 minutes
30	2 hours 32 minutes

Future work

Speed

Specificity

- Repeat experiment with heterogeneous sample

High Efficiency

- COMSOL Simulation to understand relation between Drag and DEP force
- Fabrication of electrodes

High throughput

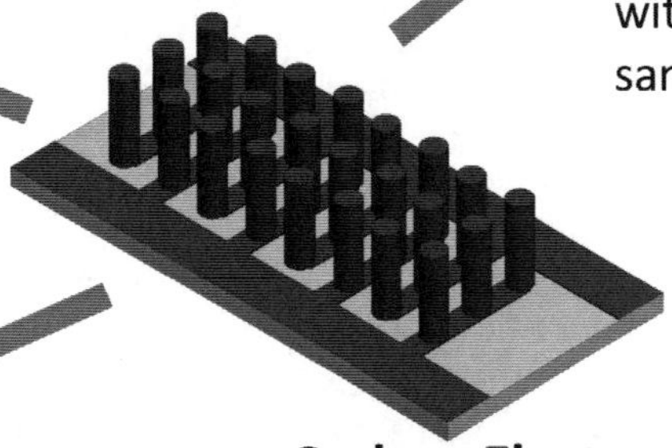

Carbon Electrode Dielectrophoresis

www.multiscalemanufacturing.net 12

Thank you

www.multiscalemanufacturing.net 13

420057 Dielectrophoretic Behavior of Pegylated Ribonuclease a in a Microdevice with Diamond-Shaped Insulating Posts

Abstract:

Ribonuclease A (RNase A) has proven potential applications as a therapeutic drug for tumor treatment, especially in its PEGylated form. Grafting of polyethylenglycol (PEG) molecules to this protein yields mono-PEGylated (mono-PEG) and di-PEGylated (di-PEG) RNase A conjugates and the remaining unreacted protein after reaction. From these species, mono-PEG RNase A has the highest biological activity. Recovery of mono-PEGylated RNAse A is a challenging task. Unlike chromatographic methods, which have been traditionally used to separate and recover mono-PEG RNase A, the use of electrokinetic foreces in microdevices represents an attractive option for this purpose. This work describes the dielectrophoretic behavior of PEGylated RNase A species and its native form inside an insulator-based dielectrophoresis (iDEP) device in route to design a microbioprocess to recover mono-PEGylated RNase A from the whole PEGylation mixture. The 10 mm long microdevice has a single row of six diamond-shaped posts embedded along the center of channel, each post having a half diamond placed on both sidewalls resulting in a device with two openings. The geometric array allowed the creation of high non-uniform electric fields when direct current electric fields were generated. As expected, the three proteins under study (i.e. native, mono-PEG, and di-PEG) exhibited different and distinct dielectrphoretic behaviors. Strong trapping of mono-PEGylated RNase A was observed when an electric potential of 4000 V was applied (Figure 1A). On the other hand, di-PEGylated RNase A showed a similar response when the electric potential applied was of 2500 V (Figure 1B). In the case of the native RNase A, no trapping was observed. The distinct dielectrophoretic behaviors of the three proteins open the possibility of designing novel devices to separate, concentrate, and recover one single PEGylated conjugate in a one step operation minimizing time, sample size, and reagent usage. This technique can serve as an alternative to traditional chromatographic methods in route to bioprocess miniaturization.

Improving the Design of Insulator-Based Dielectrophoretic Devices: Effect of Insulator Posts Characteristics

Mario Alberto Saucedo-Espinosa, Mallory Rauch, Blanca H. Lapizco-Encinas*

Microscale Bioseparations Laboratory, Rochester Institute of Technology, Rochester, NY, US

Introduction

Insulator based dielectrophoresis (iDEP) is a rapidly growing microfluidics technique, with potential applications in a wide array of fields, including environmental monitoring, bioanalytical analysis and biomedical and clinical assessments [1,2].

In iDEP particles are manipulated by dielectrophoretic forces created by the presence of insulating structures embedded in a microchannel. An electric field is usually applied employing external electrodes. The electric field is distorted by the insulating posts, creating zones of higher and lower field intensity, *i.e.*, an electric field gradient, which in turn produces dielectrophoretic forces. There are numerous reports on the successful manipulation of particles employing iDEP systems. Cell viability assessments have been performed where differences in the dielectric properties of the cell membrane can be exploited to separate live and dead cells [3-5], a recent report on the effect of iDEP manipulation is also available in the literature [6]. The manipulation of macromolecules such as DNA [7,8] and proteins [9-11] has received significant attention in the field of DEP. Sophisticated cells assessments employing simple systems with insulating posts have demonstrated the potential of iDEP for many more applications in all areas of bioanalysis. Some examples are single cells analysis and cancer cell analysis carried in iDEP systems [12,13].

Insulator- based DEP also offers attractive advantages over other DEP-based systems: simple device fabrication, potential for disposable devices and the ease of employing electroosmosis to promote fluid flow within the device, which eliminates the requirement of a micro-pump [14].

Due to these attractive characteristics, there is a significant interest in improving the design of iDEP devices. A number of reports on this can be found in the literature.

---

* Email:bhlbme@rit.edu

Kwon *et al.* [15] studied the spacing between insulating posts in order to maximize the potential for particle trapping and enrichment. Weiss *et al.* [16] studied the magnitude of the dielectrophoretic mobility of microparticles in a funnel-type channel, mobility information is important when designing this systems. Two recent reports by our group [1,17] have studied in detail the effect of the post geometry and particle shape on the performance of iDEP systems in term of particle trapping efficiency.

The present work analyzes in detail the geometrical parameters of the insulating posts (shape and spacing) with the aim of enhancing particle trapping. This study comprises extensive mathematical modeling and experimental work. The fabrication feasibility of the devices is also analyzed.

Theoretical background

The DEP velocity and force for an spherical particle of radius r_p are defined [18]:

$$\vec{v}_{dep} = \mu_{dep}\nabla\left(\vec{E}\cdot\vec{E}\right) = \frac{r_p^2\varepsilon_m}{3\eta}\mathrm{Re}\left[f_{CM}\right]\nabla\left(\vec{E}\cdot\vec{E}\right) \tag{1}$$

$$\vec{F}_{dep} = 2\pi r_p^3\varepsilon_m\,\mathrm{Re}\left[f_{CM}\right]\nabla\left(\vec{E}\cdot\vec{E}\right) \tag{2}$$

where μ_{dep} denotes the DEP mobility and $\mathrm{Re}[f_{CM}]$ is the real part of the Clausius-Mossotti (CM) factor. The CM factor accounts for the relative polarizability of the particles compared to that of the suspending medium and dictates if they exhibit positive or negative DEP. When low frequency electric fields are used, as in the case of direct current, the CM factor can be approximated by the real conductivities as described next:

$$f_{CM} = \left[\frac{\sigma_p - \sigma_m}{\sigma_p + 2\sigma_m}\right] \tag{3}$$

where σ_p and σ_m are the particle and suspending medium conductivities respectively. This study assumed that polystyrene particles exhibited negative DEP [19]. In iDEP systems, particles experience other forces besides DEP. Linear electrokinetics is defined as the superposition of electroosmosis (EO) and electrophoresis (EP). The EO and EP velocities are given by the Helmholtz-Smoluchowski equation. The expressions for the EK velocity and force for a spherical particle are defined as:

$$\vec{v}_{ek} = \mu_{ek}\vec{E} = -\frac{\varepsilon_m(\zeta_w - \zeta_p)}{\eta}\vec{E}, \qquad (4)$$

$$\vec{F}_{ek} = -6\pi r_p^2 \varepsilon_m(\zeta_w - \zeta_p)\vec{E}, \qquad (5)$$

where μ_{ek} denotes the EK mobility; defined by the real permittivity (ε_m) and viscosity (η) of the suspending medium. Additionally, ζ_w and ζ_p are the zeta potentials of the microchannel wall and particle, respectively. The microchannels, made from polydimethylsiloxane (PDMS), and the polystyrene particles used in this study have a negatively surface charge. The EK motion is towards the ground electrode (outlet reservoir, see Fig. 1).

Mathematical model in COMSOL

To perform this study, a trapping condition for particles was defined. Particles become trapped when they reach a net zero velocity, which occurs when the EK and DEP forces are in equilibrium; that is, the forces have the same magnitude but opposite directions. The expression for this trapping condition is included below in terms of velocities and mobilities [20]. When this condition is satisfied (magnitude ≤1) particles are trapped.

$$\frac{\vec{v}_{dep}\vec{E}}{\vec{v}_{ek}\vec{E}} = \frac{\mu_{dep}}{\mu_{ek}} \frac{\nabla(\vec{E}\cdot\vec{E})\cdot\vec{E}}{\vec{E}\cdot\vec{E}} \le -1. \qquad (6)$$

Experimental section

Microdevices

Microdevices were fabricated from PDMS employing standard soft lithography techniques. Each microchannel was 10.16 mm long, 1 mm wide and 40 µm deep, and contained an inlet and outlet liquid reservoirs. The array of insulator posts was located at the channel center, 4 mm away from the inlet reservoir and patterned in arrays of 16 columns of 4 posts each (Figure 1). The shape and spacing of the insulating post was varied and different devices were used.

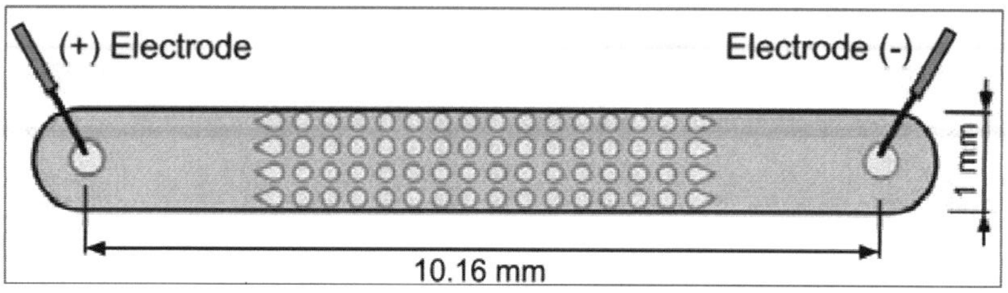

Figure 1. Schematic representation of the initial microchannel employed in this study.

Microparticles

Yellow-green fluorescent polystyrene microparticles (ex/em 505/515 nm) with a diameter of 2-μm were employed for experimentation. Particles were suspended in DI water with pH = 8 and σ_m = 20 μS/cm to a concentration of 7.0×10^8 particles/mL.

Results and discussion

The insulating posts distort the electric field distribution in the channel, therefore the spacing and shape of the insulating posts has an important impact of the dielectrophoretic forces generated within the channel. For this study, we selected as the region of interest the spacing between four posts. To analyze the distribution of the particle velocity in this region, we divided the region in two zones: upstream and downstream. As it can be seen from Figure 2, the direction of the net particle velocity changes within the region. It can also be observed that the particle velocity is asymmetric through this region and that there is a location where the particle velocity is zero. In the upstream zone the particle velocity is high since DEP and EK forces are pushing in the same direction. In contrast, in the downstream direction, EK and DEP forces have an opposite direction, where DEP dominates, leading to a lower magnitude in the net particle velocity (as indicated by the absence of red color in the downstream position). The spacing between consecutive posts plays a major role in this velocity distribution and this spacing can be optimized to maximize particle trapping. Kwon *et al.* [15] optimized particle trapping considering positive DEP. Our work is focused on maximizing particle trapping considering negative DEP.

To further analyze the effects of post shape, we studied the distribution of the electric field ($\vec{E}$) and the gradient of the magnitude of the electric field squared (∇E^2) for two different post shapes: diamond-shaped posts and cylindrical posts. Figure 3 illustrates these results, where it can be seen that the cylindrical posts (Figs. 3d-3f) offer a higher electric field maxima and also a slightly higher maxima of ∇E^2 than the diamond-shaped posts (Figs. 3a-3c). Furthermore, the cylindrical posts generate larger regions of higher ∇E^2 than the diamond-shaped posts, which is an important advantage for particle trapping. A high value in the magnitude of ∇E^2 over a small region may not produce effective particle trapping, the size of the region influenced by ∇E^2 directly affects the capacity for particle trapping.

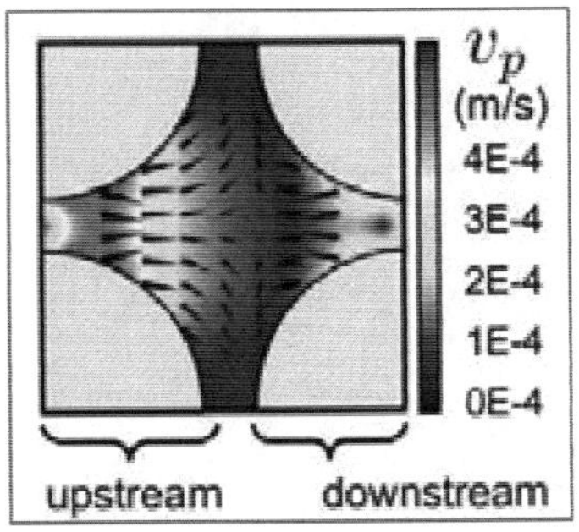

Figure 2. Particle velocity magnitude (v_p) in the opening obtained applying a voltage of 1400 V conspiring cylindrical posts with a diameter of 200 μm and spaced 250 μm center-to-center. The downstream and upstream regions represent the zones where particle velocity increases and decreases, respectively, as a result of the negative DEP force.

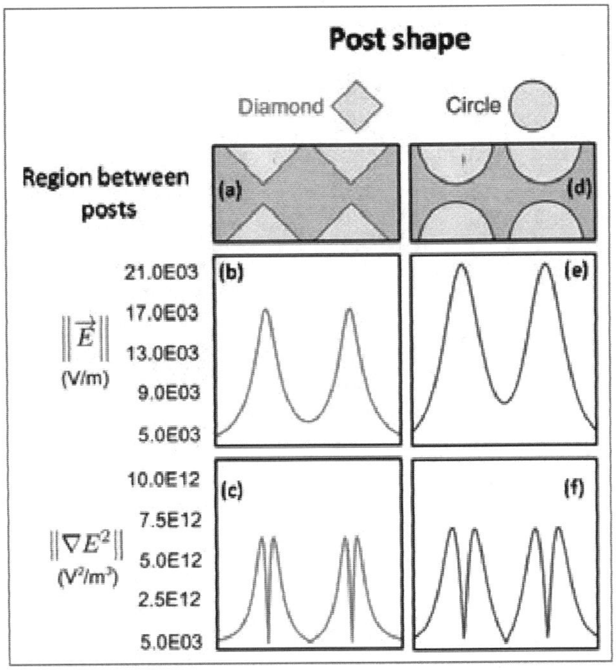

Figure 3. Region between the posts, distribution of the electric field and gradient of the magnitude of the electric field squared for diamond-shaped posts (a-c) and cylindrical posts (d-f). These values were obtained considering a potential of 1400 V and post diameter/width of 200 μm and 250 μm center-to-center spacing.

To further analyze the effect of post shape, a series of experiments were carried out. A recent report by our groups has already remonstrated that narrow posts produce higher trapping [17]. The results in Figure 3 illustrated that cylindrical or rounded geometries have the capacity of generating larger zones for particle trapping. By combining these findings, we designed oval-shaped post that had excellent performance as shown in Figure 4. The final shape of these optimized posts was designed employing mathematical modeling; numerous simulations were run to perform this optimization. After the final designs were decided the devices were fabricated and tested employing 2-μm polystyrene particles. We carried experiments with "narrow diamonds" and "narrow circles" (ovals). Figure 4a shows significant particle trapping obtained with narrow diamond-shaped posts at an applied potential of 450 V. However, superior particle trapping at only 170 V was obtained with the oval-shaped post as illustrated in Figure 4b.

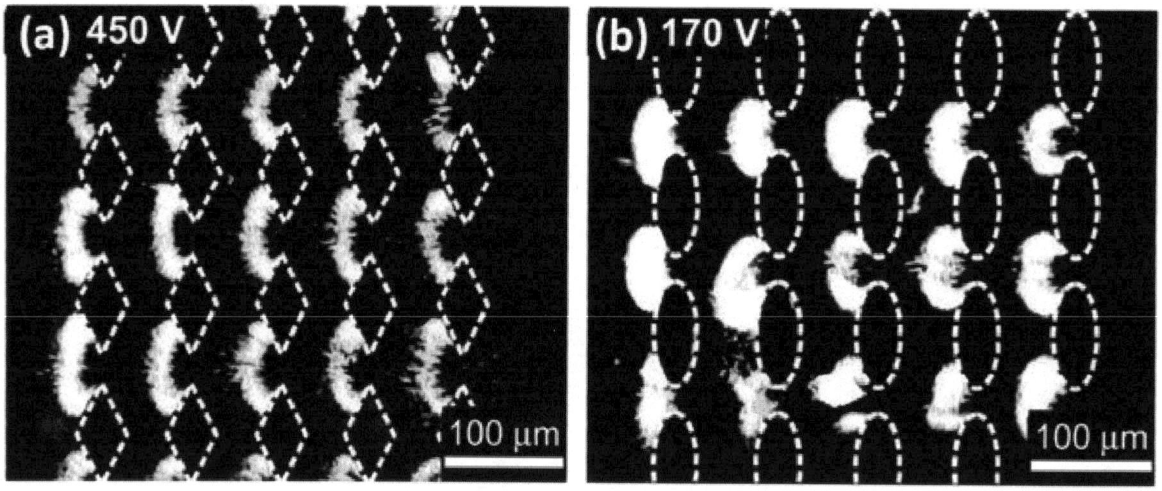

Figure 4. Experimental bands of trapped 2-μm particles employing (a) narrow diamond-shaped posts with a width of 48 μm and height of 80 μm, spaced 20 μm horizontally and 8 μm vertically; and (b) oval shaped posts width a width of 32 μm and height of 80 μm, spaced 36 μm horizontally and 8 μm vertically. This figure clearly illustrates a much better performance of the oval-shaped posts obtained at the low applied potential of 170 V.

Conclusions

This study demonstrates the importance of the shape and spacing of the posts insulator-based DEP (iDEP) microdevices. Numerical simulations in COMSOL Multiphysics® were employed to find the best designs that maximize particle trapping performance. These designs were tested experimentally employing polystyrene microparticles. It was found that oval-shaped insulating post produce the best performance while requiring lower applied electric potentials. The findings from this study provide a systematic approach towards the design of improved high-performance iDEP devices.

Acknowledgments

The authors would like to acknowledge the financial support provided by the National Science Foundation (Award CBET-1336160).

References

[1] Saucedo-Espinosa, M. A., Rauch, M., Lapizco-Encinas, B. H., in *AICHE Annual Meeting 2014*, AICHE, Atlanta GA, USA, November 2014, 2014.

[2] Regtmeier, J., Eichhorn, R., Viefhues, M., Bogunovic, L., Anselmetti, D., *Electrophoresis* 2011, *32,* 2253-2273.

[3] Patel, S., Showers, D., Vedantam, P., Tzeng, T.-R., Qian, S., Xuan, X., *Biomicrofluidics* 2012, *6,* 034102-034112.

[4] Gallo-Villanueva, R. C., Jesús-Pérez, N. M., Martínez-López, J. I., Pacheco, A., Lapizco-Encinas, B. H., *Microfluid. Nanofluid.* 2011, *10,* 1305-1315.

[5] Lapizco-Encinas, B. H., Simmons, B. A., Cummings, E. B., Fintschenko, Y., *Anal. Chem.* 2004, *76,* 1571-1579.

[6] LaLonde, A., Romero-Creel, M. F., Lapizco-Encinas, B. H., *Electrophoresis* 2014, *36,* 1479-1484.

[7] Camacho-Alanis, F., Gan, L., Ros, A., *Sens. Actuator B-Chem.* 2012, *173,* 668-675.

[8] Kuzyk, A., *Electrophoresis* 2011, *32,* 2307-2313.

[9] Abdallah, B. G., Chao, T.-C., Kupitz, C., Fromme, P., Ros, A., *ACS Nano* 2013, *7,* 9129-9137.

[10] Chaurey, V., Rohani, A., Su, Y.-H., Liao, K.-T., Chou, C.-F., Swami, N. S., *Electrophoresis* 2013, *34,* 1097-1104.

[11] Nakano, A., Camacho-Alanis, F., Chao, T.-C., Ros, A., *Biomicrofluidics* 2012, *6,* 034108-034113.

[12] Bhattacharya, S., Chao, T.-C., Ariyasinghe, N., Ruiz, Y., Lake, D., Ros, R., *et al.*, *Anal. Bioanal. Chem.* 2014, *406,* 1855-1865.

[13] Bhattacharya, S., Chao, T.-C., Ros, A., *Electrophoresis* 2011, *32,* 2550-2558.

[14] Srivastava, S., Gencoglu, A., Minerick, A., *Anal. Bioanal. Chem.* 2010, *399,* 301-321.

[15] Kwon, J.-S., Maeng, J.-S., Chun, M.-S., Song, S., *Microfluid. Nanofluid.* 2008, *5,* 23-31.

[16] Weiss, N. G., Jones, P. V., Mahanti, P., Chen, K. P., Taylor, T. J., Hayes, M. A., *Electrophoresis* 2011, *32,* 2292-2297.

[17] LaLonde, A., Gencoglu, A., Romero-Creel, M. F., Koppula, K. S., Lapizco-Encinas, B. H., *J. Chromatogr. A* 2014, *1344,* 99-108.

[18] Jones, T. B., Electromechanics of Particles, Cambridge University Press, USA, 1995.

[19] Baylon-Cardiel, J. L., Lapizco-Encinas, B. H., Reyes-Betanzo, C., Chávez-Santoscoy, A. V., Martínez Chapa, S. O., *Lab Chip* 2009, *9,* 2896-2901.

[20] Saucedo-Espinosa, M. A., Lapizco-Encinas, B. H., *Electrophoresis* 2015, *36,* 1086-1097.

DEP Isolation and Detection of cf-DNA, cf-RNA and Exosome Biomarkers from Hematological Cancer, Solid Tumor and TBI Patient Blood and Plasma Samples

Michael J. Heller, Stuart Ibsen and Jennifer Wright,
University of California San Diego, Department of Bioengineering, Department of Nanoengineering, La Jolla, CA, USA 92093-0448

The analysis of cell free (cf) DNA and cf-RNA continue to become more widely used for cancer and other disease diagnostics and management. We have now demonstrated a dielectrophoretic (DEP) based approach that allows cf-DNA and cf-RNA biomarkers from chronic lymphocytic leukemia (CLL), solid tumor and traumatic brain injury (TBI) patients to be isolated in about 10-20 minutes directly from a small volume (25-50ul) of patient whole blood or plasma. Figure 1 shows fluorescent image analysis results for cf-DNA isolated by DEP from one normal blood sample and five CLL patient blood samples (left side), and one normal plasma sample and five CLL patient plasma samples (right side). Overall, the fluorescent cf-DNA levels were higher in most of the CLL patient samples when compared to the fluorescent DNA levels obtained from the normal blood samples. The PCR and DNA sequencing results for cf-DNA isolated from 5 CLL patient blood and plasma samples by the DEP process were exactly comparable to results obtained using conventional methods, which take several hours to complete, are more laborious (many steps) and require much larger plasma/blood samples. In the case of a solid tumors Figure 2 below shows cf-DNA which was isolated by DEP from a 50ul plasma sample from an NSC lung cancer patient. Generally, cf-DNA for "liquid biopsies" is isolated from plasma using relatively long and involved processes which are not suitable for point of care (POC) diagnostics. DEP now allows the isolation and fluorescent detection to now be carried out in less than 20 minutes. In further work related to brain cancer and TBI diagnostics, Figure 3shows the isolation of cf-RNA containing glioblastoma exosomes from human plasma samples. The new DEP devices are now being used to collect cf-DNA/RNA from a variety of other cancer (solid tumor) and TBI patient blood and plasma samples, setting the stage for "seamless sample to answer" diagnostics. *(1) Sonnenberg A, Marciniak JY, Rassenti L, Ghia EM, Skowronski EA, Manouchehri S, McCanna JP, Widhopf II GF, Kipps TJ and Heller MJ, "Dielectrophoretic Isolation and Detection of Cancer Related Circulating Cell Free DNA Biomarkers from Blood and Plasma", Electrophoresis, V35, 12-13, pp. 1828-36, July 2014. (2) Sonnenberg A, Marciniak JY, Rassenti L, Ghia EM, Skowronski EA, Manouchehri S, McCanna JP, Widhopf II GF, Kipps TJ and Heller MJ, "Rapid Electrokinetic Isolation of Cancer-Related Circulating Cell Free DNA Directly from Blood", Clinical Chemistry, 60:3, pp.500-509, 2014; (4) Lewis J, Heineck D and Heller MJ, "Detecting Cancer Biomarkers in Blood: Challenges for new molecular diagnostic and point-of-care tests using cell-free nucleic acids", Expert Review Molecular Diagnostics, Early online, 1-14, 2015.*

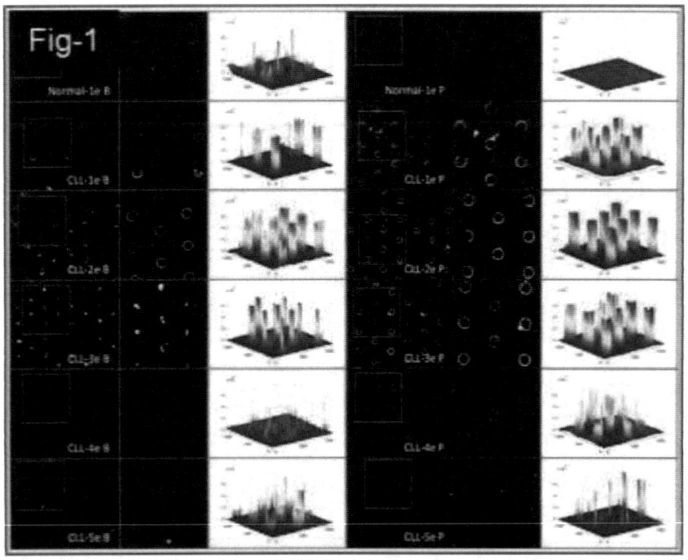

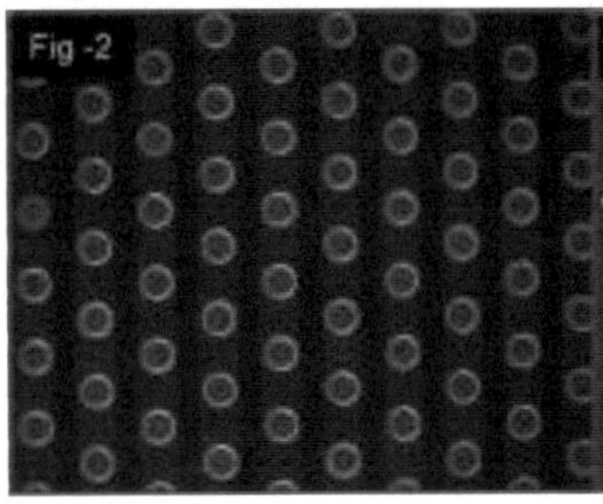

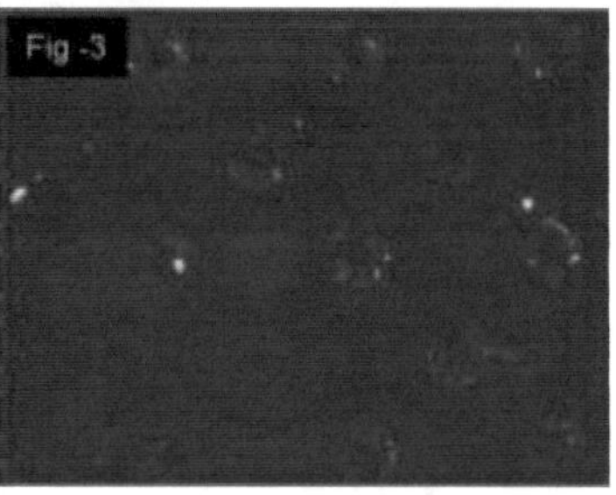

430561 Actuated Tweezers for Electrokinetic Sample Separation

Abstract:

The isolation of specific particle by means of filtration is a crucial step towards the analysis of target species within a heterogeneous input sample. There is a need for a method of targeted filtration that minimizes strain on the sample particle. Currently, conventional methods of separation and filtration, such as centrifugation and fluidic techniques, can exert potentially harmful forces on the sample resulting in the loss or damage of desired species. To address these concerns a device has been developed that can perform "stress-free" separation on a given sample. This Electrokinetic Tweezer's device utilizes a precise rail system to actuate electrode "tweezers" through a sample to selectively collect targets for removal. The device utilizes the phenomenon of dielectrophoresis to allow for selective collection by applying a non-uniform sinusoidal electric field ranging between 0 – 10 Vpp and 0 – 10 MHz. The result is an applied force on the desired particles drawing them towards the electrodes for removal. As opposed to traditional dielectrophoretic separation techniques the electrokinetic tweezers allow for a stationary sample with actuated electrodes eliminating the stress attributed to the motion associated with centrifugal forces or fluidics. The device was tested on a sample of red blood cell phantoms with comparable dielectric properties to red blood cells in human blood plasma and demonstrated a rate of removal near 200,000 targets/min. The desktop device is versatile with a modular design allowing for the exchange of different electrode designs, channel geometries, and applied electric field. It also easily interfaces with standard infusion pumps for automated sample loading. The electrokinetic tweezer's shows promise in a variety of fields such as plasmapheresis and cell sorting giving it the potential to become a low-cost and robust multipurpose clinical tool.

417166 Mathematical Modeling of Tubulin Transport in Tumor Cells Under the Influence of Electrical Fields

Abstract:

Recently, low magnitude, intermediate frequency alternating current (AC) applied electrical fields have been shown to disrupt cellular mitosis, having huge implications for cancer therapy. It has been hypothesized that these fields inhibit tubulin polymerization, preventing the formation of microtubules, thus arresting tumor cell division. Mathematical models based on fundamental principles can be used to gain a mechanistic understanding of tumor cell response to these fields. As a first step in understanding this behavior, we will present a preliminary model that couples tubulin transport to fluid motion within the cell when DC and AC electrical fields are applied. This is accomplished through the use of the electrokinetic-hydrodynamics (EKHD) approach to compute effective transport coefficients (effective velocity and diffusivity) of tubulin within tumor cells undergoing mitosis. These results will be compared to limiting cases in which there is no electrical field applied and when the cells are not undergoing mitosis.

Dielectrophoretic Isolation and Enrichment of Low Abundant Particles

Alexandra LaLonde, Maria F. Romero-Creel, Mario A. Saucedo-Espinosa and Blanca H. Lapizco-Encinas*

Microscale Bioseparations Laboratory, Rochester Institute of Technology, Rochester NY, US.

Introduction

The detection of low-abundant particles is of interest when considering bio-analytical and clinical applications; for example when detecting pathogens in blood, and other contaminants like circulating tumor cells (CTCs). There is an increase of interest in the development of microfluidic techniques that possess screening capabilities accurate enough to detect low-concentrated particles and isolate them from complex solutions [1-3]. Due to the portability, fast processing time, and low cost, microfluidics has transformed the way that many analytical assessments are performed. Dielectrophoresis (DEP) and electrophoresis (EP) are examples of electric field driven techniques that have proven to be reliable for assessing cells and inert particles. DEP is particularly attractive due to its flexibility allowing for separation and enrichment of particles by exploiting polarization effects with either alternating current (AC) or direct current (DC) electric fields. There are two major classifications of dielectrophoresis: electrode-based DEP (eDEP) and insulator-based DEP (iDEP). IDEP systems offer less complex devices compared to eDEP systems. IDEP devices are quick and inexpensive to fabricate due to the entire device usually being made from a single substrate. As an additional advantage, iDEP systems utilize electroosmotic (EO) flow as means for pumping liquid and particles through the microchannels, simplifying these systems further. As a downfall these systems do however require much greater electric potentials in order to produce high enough dielectrophoretic forces.

* Email:bhlbme@rit.edu

Several studies using microfluidic devices for diagnostics and clinical applications have been significantly explored using DEP-based methods. In particular there are important reports on the use of eDEP systems for detection and purification of rare cells. Gascoyne *et al.* [2] employed a field-flow-fractionation DEP system and reported the isolation of CTCs from blood. A DEP device with embedded microelectrodes on the bottom of the channel was used to analyze concentration ratios of 1:1000 CTCs to normal blood cells. Tumor cells were successfully separated from normal cells by exploiting negative DEP behavior. The potential of this eDEP system for rare cell isolation was demonstrated as recovery rates from 48-92% were obtained. In another study, Kuczenski *et al.* [3] isolated *Escherichia coli* cells from blood using a system with patterned microelectrodes. Samples containing a concentration ratio 1:1 of *E. coli* cells to red blood cells (RBC) were employed. As a result bacterial cells were enriched (five-fold increase) and separated from RBC; thus demonstrating the potential of DEP for pathogen detection.

As depicted above current studies require use of microelectrodes to power their systems and external pumps to generate liquid flow. IDEP systems would offer simplicity and ease in order to carry out these same isolations and enrichments of particles. A full characterization of the limits and constraints these iDEP systems have for handling low abundant particles has not yet been constructed. The present study analyzes the potential for a simplistic iDEP system to detect, capture, and enrich low abundant cells and particles in a mixture. . Binary and tertiary mixtures of yeast cells and polystyrene particles were analyzed, where yeast cells were the low concentrated particles in the mixture. The mixtures studied varied in concentration ratios with orders of 1:10,000,000 yeast cells compared to the other particles present in the mixture. The effectiveness of iDEP was evaluated by varying the particle concentration ratio, and particle mixture concentration. The results of this study are promising, since it was demonstrated that simple iDEP systems can be used for detecting, and capturing significantly low-concentrated particles from a mixture.

Theory of dielectrophoresis

A non-uniform electric field is needed for dielectrophoresis to occur. DEP depends on the electric field gradient (∇E^2), not the electric field itself. Under a DC electric field the dielectrophoretic force exerted on a spherical particle is defined as [4]:

$$\vec{F}_{DEP} = 2\pi\varepsilon_m r_p^{\,3}\,\mathrm{Re}(f_{CM})\,\nabla(\vec{E}\cdot\vec{E}) \tag{1}$$

where ε_m is the permittivity of the suspending medium, r_p is the radius of the particle, $\nabla(\vec{E}\cdot\vec{E})$ is the squared gradient of the field strength, and Re(f_{CM}) is the real part of the Clausius-Mossotti factor (f_{CM}). Particle polarizability relative to that of the suspending medium is accounted for by f_{CM}. Under DC and low frequency AC-electric fields, the f_{CM} can be expressed in terms of the real conductivities of the particles (σ_p) and the suspending medium (σ_m) [5,6]:

$$f_{CM} = \frac{\sigma_p - \sigma_m}{\sigma_p + 2\sigma_m} \tag{2}$$

Other than DEP, electrophoresis (EP) and EO flow are the two significant forces present in DC-iDEP systems. The superposition of EO flow and EP is known as the electrokinetic (EK) force. The expression for the EK velocity ($\vec{v}_{EK}$) in terms of zeta potentials and mobilities is [7]:

$$\vec{v}_{EK} = \mu_{EK}\vec{E} = (\mu_{EOF} + \mu_{EP})\vec{E} = \frac{\varepsilon_m}{\eta}(-\zeta_{wall} + \zeta_p)\vec{E} \tag{3}$$

where μ_{EK}, μ_{EO} and μ_{EP} are the EK, EO and EK mobilities, respectively; η is the suspending medium viscosity and ζ_{wall} and ζ_p are the zeta potentials of the microchannel wall and the particle, respectively. The devices for this study were made from polydimethylsiloxane (PDMS) which has a negative zeta potential. As a result EO flow is produced from the positive to the negative electrode (left to right) by applying DC potentials.

Experimental section

Sample preparation

Fluorescent polystyrene microspheres sizes (Invitrogen, Eugene, OR) of three different sizes were employed: 500-nm diameter green (ex/em 505/515 nm) and red (ex/em 580/605 nm), and 1-μm diameter yellow-green (ex/em 505/515 nm). Stock suspensions of these microspheres were vortexed in order to disrupt aggregates and then diluted in suspending medium to the desired particle concentrations.

Saccharomyces cerevisiae cells (ATCC 9763, Manassas, VA, USA), with an average diameter measured as 6.3 ± 0.4 μm, were cultured in YM medium (Amresco, Albany, NY, Cat. 271120) at 30°C in a shaking incubator for 16h. Once the concentration of cells was determined to be ~8.37×10^5 cells, a volume of 1.75 mL of culture was centrifuged at 10,000rpm for 15 minutes and the supernatant was then discarded. The left over pellet was then washed by re-suspending it in 1.75 mL of DI water and centrifuged again at 10,000rpm for 15 minutes and discarding the supernatant. The pellet was re-suspended in 1.75 mL of DI water and 6 μL of Syto 11(Invitrogen, Eugene, OR) green fluorescent (ex/em 508/527 nm) nucleic stain was added. The tube containing the sample was then covered with aluminum foil and incubated for 40 minutes at room temperature. This sample was then washed and a final volume of 1.1 mL of DI water was added to the residual pellet, resulting in a final concentration of 1×10^7 cells/mL.

Microdevice Description

A microchannel fabricated using PDMS (Dow Corning, Midland, MI) and standard soft lithography techniques [8], was used for DEP experiments employing circular post shapes 200-μm in diameter and arranged 250-μm center to center. The channel was 10.16-mm long and 1-mm wide, with a depth of 40-μm containing one inlet and one outlet liquid reservoirs for results obtained in Figure 2 and Figure 3. A schematic representation of this microdevice is shown in Figure 1.

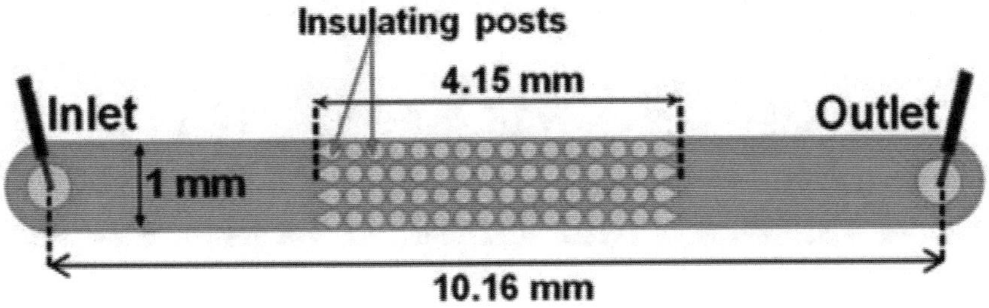

Figure 1. Schematic representation of the iDEP channel employed for the experiments shown in Fig. 2 and Fig. 3.

Equipment and Software

A Trek high voltage amplifier (Model: PZD700A2, Trek Inc., Lockport, NY) and an Agilent waveform generator (Agilent 33500 Series, Santa Clara, CA) were used to apply DC electric potentials for sections XXX. For sections XXX, a high voltage sequencer (Model HVS6000D; LabSmith, Livermore, CA) was used. For all experiments, platinum wire electrodes were used to apply the electric potentials. All iDEP experiments were carried out using a ZEISS Axiovert 40 CFL inverted microscope (Carl Zeiss Microscopy, Thornwood, NY), and an Infinity 2 camera (Luminera, Ottawa, Canada) was used to visualize the experiments.

Results and discussion

One of the main objectives of the study was to test the limits of the simple iDEP technique, selectively capturing and enriching low-abundant particles in a mixture. Particle ratios as large as 1:10,000,000 were employed, where the large particle in the mixture was defined to be the low-abundant particle. It can be seen in Figure 2, that iDEP devices can strongly capture these low-concentrated particles in systems where other EK forces (EP and EO flow) and other particles are present. It was also important for this study to test the performance of the device employing biological cells as the low-abundant cells, in order to determine its potential application for manipulation and processing of biological samples. Figures 2a-2d depict selective and reliable iDEP trapping of yeast cells in a mixture with 500-nm particles with concentration ratios up to 1:10,000,000.

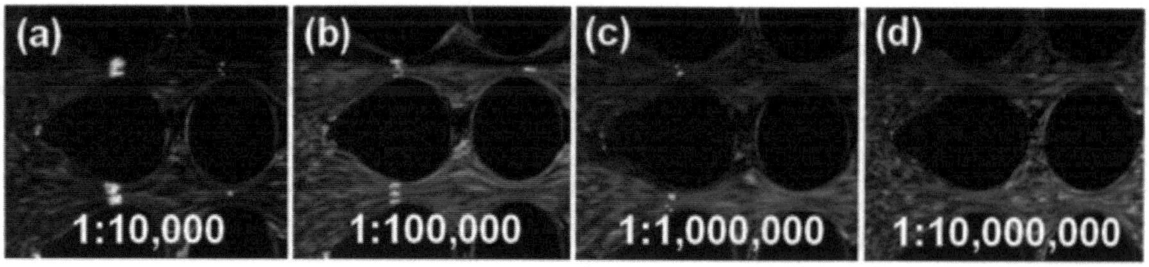

Figure 2. Experimental trapping of *yeast* (green) cells in a mixture with 500-nm (red) particles obtained at 600 V. The concentration of 500-nm was kept constant at $1x10^9$ particles/mL. The following yeast cell concentrations were used in cells/mL: (a) $1x10^5$, (b) $1x10^4$, and (c) $1x10^3$, (d) $1x10^2$. Concertation ratios are depicted in each image.

Tertiary particles were employed to further test the limits of iDEP devices. Yeast fluorescent cells were used as the low-abundant particles in a mixture with 1-μm and 500-nm polystyrene particles with a concentration ratio of 1:1,000:10,000, respectively. Figure 3 illustrates the successful trapping of yeast cells achieved with this mixture, demonstrating that iDEP systems offer a reliable alternative to techniques used for analysis of low concentrated particles of interest.

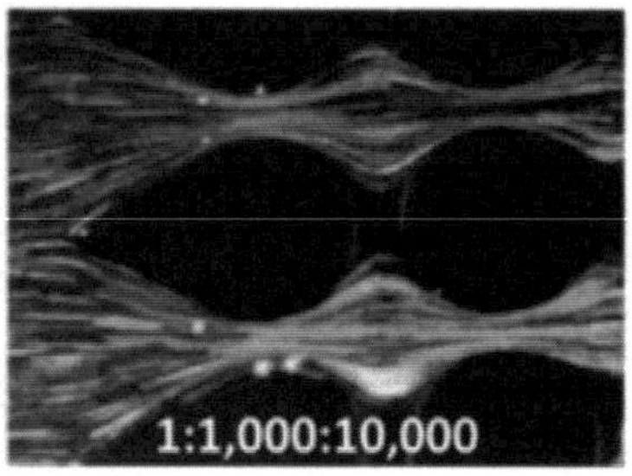

Figure 3. Experimental trapping of *yeast* (green) cells in a mixture with 1-μm (red) and 500-nm (green) polystyrene particles, employing a concentration ratio of 1:1,000:10,000, respectively. The concentrations and applied voltages employed were $1x10^5$: $1x10^8$: $1x10^9$ at 350 V

Conclusion

Analysis of samples containing low-abundant particles is an important application in various fields. The present work tested the potential of simple iDEP devices with circular posts as a technique for capturing and enriching such rare particles in binary and tertiary mixtures. Different samples containing yeast cells,1-μm, and 500-nm polystyrene particles were used (in ratios up to 1:10,000,000) where iDEP systems successfully demonstrated dielectrophoretic trapping of the low-abundant particles. These results are encouraging, since they demonstrate that effective detection and enrichment of rare particles can be performed with simple iDEP systems, opening the possibility for iDEP as a technique of choice for this type of analysis.

Acknowledgements

The authors would like to acknowledge the financial support provided by the National Science Foundation (Award CBET-1336160).

References

[1] Huang, C.-T., Amstislavskaya, T. G., Chen, G.-H., Chang, H.-H., Chen, Y.-H., Jen, C.-P., *J. Med. Biol. Engg.* 2013, *33,* 51-58.

[2] Gascoyne, P. R. C., Noshari, J., Anderson, T. J., Becker, F. F., *Electrophoresis* 2009, *30,* 1388-1398.

[3] Kuczenski, R. S., Chang, H.-C., Revzin, A., *Biomicrofluidics* 2011, *5,* 032005-032015.

[4] Kirby, B. J., Micro- and Nanoscale Fluid Mechanics. Transport in Microfluidic Devices, Cambridge University Press New York, 2010.

[5] Jones, T. B., Electromechanics of Particles, Cambridge University Press, USA, 1995.

[6] Markx, G. H., Dyda, P. A., Pethig, R., *J. Biotechnol.* 1996, *51,* 175-180.

[7] Moncada-Hernandez, H., Baylon-Cardiel, J. L., Pérez-González, V. H., Lapizco-Encinas, B. H., *Electrophoresis* 2011, *32,* 2502-2511.

[8] Duffy, D. C., McDonald, J. C., Schueller, O. J. A., Whitesides, G. M., *Anal. Chem.* 1998, *70,* 4974-4984.

Reversing elution order in insulator based dielectrophoretic separations with asymmetric insulating posts

Mario A. Saucedo-Espinosa, Alexandra LaLonde, Aytug Gencoglu, María F. Romero-Creel, Jay Dolas and Blanca H. Lapizco-Encinas*

Microscale Bioseparations Laboratory, Rochester Institute of Technology, Rochester NY, US

Introduction

Dielectrophoresis (DEP) refers to the transport mechanism caused by polarization effects when a particle is exposed to a non-uniform electric field [1,2]. The nature of the DEP force exerted on a particle is a function of the relative properties of the particle and suspending medium. In the case where the particle has a higher polarizability than that of the suspending medium, the particle exhibits positive DEP, and move towards the regions with higher electric field gradient. Negative DEP occurs when particles are repelled from these regions [1,2].

In insulator-based DEP (iDEP), the presence of insulating structures distort the electric field distribution and cause regions with a non-uniform electric field. Electrophoresis (EP) and electroosmotic (EO) flow effects are significant in iDEP systems, since high direct current (DC) or low frequency alternate current (AC) potentials are used. Electrokinetics (EK), which has a linear dependence on the electric field, comprises the superposition of EO flow and EP [3,4].

The separation and enrichment of particles is a main application of iDEP systems. In particular, bioanalytical and clinical assessments are particularly attractive. Using iDEP systems, several research groups have enriched and separated bacteria mixtures [4-6], live and dead bacteria [7], human white blood cells and breast cancer cells [8]. In addition to cell manipulation, iDEP has also been applied for the manipulation of proteins [9-11], DNA [12,13], organelles [14] and virus [15,16].

In a conventional iDEP device, all particles in the sample are enriched with negative DEP. This is possible by applying DC or low frequency AC electric potentials [17,18,6]. The smaller particles are released as a "peak" of concentrated particles by lowering the applied potential. Larger particles are subsequently released by further lowering

* Email:bhlbme@rit.edu

the potential. A main disadvantage of this approach is that larger particles are exposed to electric fields for longer times. This can be problematic when sensible biological particles are present in the sample. For example, a prolonged exposure of eukaryotic cells to electric fields can lead to significant cell damage [19-21]. A recent study from our group [19] showed that only 5% of yeast cells remained viable after a 4-minute exposure when a DC potential of 1,000 V was applied in an iDEP device. All these studies have demonstrated that high electric field magnitudes and/or long exposures can lead to significant cell damage.

This work presents a system where larger particles are eluted first. This is counter-intuitive for low frequency AC-iDEP and DC-iDEP devices [6,22,17]. By employing a unique asymmetric post geometry (Fig. 1) coupled with DC-biased AC electric potentials, the elution order was reversed. A mixture of 500-nm and 1-μm polystyrene particles was separated using this technique, showing the capability to enrich and separate particles by exploiting relatively small differences in particle size. Further, a second mixture containing yeast cells (6.3-μm) and polystyrene particles (2-μm) was studied. The larger yeast cells were enriched and eluted in less than 40 seconds, before the smaller polystyrene particles. In addition, the technique added the plus of preserving the great majority of the cells viable (viability ~ 85%).

Theoretical background

In DC-iDEP and low frequency AC-iDEP devices, EO flow and EP affect particle movement, both being linear functions of the electric field. Hence, their contribution to particle movement is considered as a single EK transport term. The EK force exerted on a spherical particle of radius r_p is defined as [1]:

$$F_{EK} = 2\pi r_p \eta \vec{u}, \tag{1}$$

where η and $\vec{u}$ are the viscosity and velocity of the suspending medium in the vicinity of the particle, respectively. The DEP force exerted on a spherical particle is defined as [1,2]:

$$F_{DEP} = 2\pi \varepsilon_m r_p^3 f_{CM} \nabla(\vec{E}^2), \tag{2}$$

where ε_m is the permittivity of the suspending medium and f_{CM} is the Clausius-Mossotti (CM) factor. All particles and cells under the experimental conditions of this work exhibited negative DEP behavior.

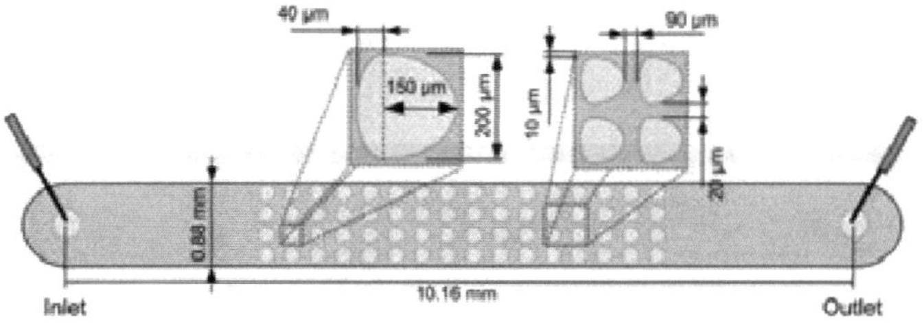

Figure 1: Schematic of the channel with array of asymmetric insulating posts.

Experimental section

Polystyrene particles

Fluorescent 500-nm (ex/em 580/605 nm), 1-µm (ex/em 505/515 nm) and 2-µm (ex/em 580/605 nm) polystyrene particles from Invitrogen (Carlsbad, CA) were employed for experimentation. The stock suspensions of the microspheres where diluted in 0.2 mM K_2HPO_4 (pH = 7.21, conductivity = 59.3 µS/cm) to final concentrations of 7.28×10^8, 7.28×10^8 and 1.03×10^8 particles/mL, respectively.

Cells

Saccharomyces cerevisiae (ATCC 9763) cells were cultured in a yeast mannitol broth liquid medium at 30 °C in a shaking incubator for 12-14 hours. For labelling, 2 mL of cell culture were centrifuged at 8,000 rpm for 5 minutes and the supernatant was discarded. The cell pellet was washed with DI water and resuspended in 0.5 mL of DI water. Next, 3 µL of Syto 11 (ex/em 508/527, Invitrogen, Carlsbad, CA) were added. Cells were then incubated with the dye for 40 minutes and centrifuged at 8,000 rpm for 5 minutes. The pellet was washed with DI water to remove the excess of Syto 11. Finally, labeled cells were resuspended in 0.252 mL of 0.2 mM K_2HPO_4. The final concentration was 1.0×10^8 cells/mL. *S. cerevisiae* cells had a semi-spherical shape with a diameter of 6.3±0.4 µm.

Equipment

A waveform generator (33522A, Agilent Technologies, Santa Clara, CA) and a high voltage amplifier (PZD700A2, Trek Inc., Lockport, NY) were used to apply AC and DC electric potentials. Images and videos of the experiments were captured with an Axiovert 40 CFL microscope (Carl Zeiss Microscopy, Thornwood, NY) equipped with an Infinity 2 camera (Luminera, Ottawa, Canada). The post-processing of the images and videos, including fluorescence signal acquisition, was performed using Fiji [23].

Microdevices

Microchannels were made from polydimethylsiloxane (PDMS, Dow Corning, Midland, MI) and fabricated using standard soft-lithography techniques [24]. Each microchannel was 10.16-mm long, 0.88-mm wide and 40-µm deep, and have an inlet and outlet liquid reservoir (Fig. 1). The asymmetric insulating posts were patterned in arrays of 16 columns with four posts each, located at the center of the microchannel. The selection of this design was based on a previous study of our group where the geometrical properties of a set of insulating posts were characterized [25].

Results and discussion

Enrichment and separation of the 1-µm and 500-nm and particles mixture

The separation of the 500-nm and 1-µm particle mixture was analyzed employing a customized set of DC-biased AC electric signals and DC potentials (Fig. 2a). There are four main steps: the elution of the 1-µm particles (Fig. 2b), their release as a "peak" of concentrated particles (Fig. 2c), the elution of the 500-nm particles (Fig. 2d) and their release as a "peak" of concentrated particles (Fig. 2e). A squared signal with a positive peak ($+V_p$) of 450 V, negative peak ($-V_p$) of -600 V and frequency (ω) of 450 mHz was used to elude and enrich the 1-µm particles. Given the smaller size for the 500-nm particles, the signal was adjusted to $+V_p$ = 1,000 V, $-V_p$ = -1,200 V and ω = 850 mHz for their elution and enrichment. Both particles were released as peaks of enriched particles by applying a DC signal of 40 V.

Figures 2b-2e show the corrected total fluorescence (CTF) signals vs. time for each consecutive step. At the beginning of the experiment, the 1-µm particles progressively concentrate at the outlet of the post array. This is noted by the steep increase in the green CTF signal (Fig. 2b). The 500-nm particles, on the other hand, keep moving towards the inlet, producing particle depletion (Fig. 2b). The enrichment of 1-µm particles and the depletion of 500-nm particles reached a limit after 70 s (Fig. 2b). When t = 120 seconds, the applied signal was switched to a DC potential for 30 seconds. This caused the EK migration toward the channel outlet of all particles. The peak of enriched 1-μm particles is shown in Figure 2c, showing a well-defined Gaussian shape with relatively small variance. This demonstrated a successful enrichment and release of the 1-μm particles. Also, the region shows a region poor in 500-nm particles, as observed in the low magnitude of the red CTF signal (Fig. 2c).

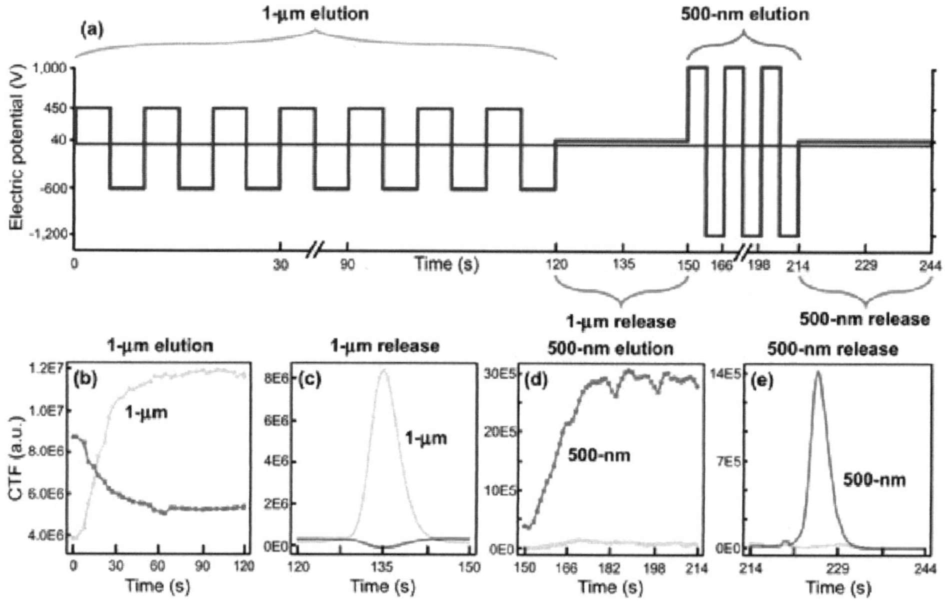

Figure 2: (a) DC-biased AC and DC electric potentials applied during the four steps of the experiments for the 1-μm and 500-nm polystyrene particles mixture. **(b-e)** CTF of the 500-nm and 1-μm particles during the **(b)** 1-μm particles elution, **(c)** 1-μm particles release, **(d)** 500-nm particles elution and **(e)** 500-nm particles release.

After the elution and release of the 1-μm particles, the applied electric potential was switched to a DC-biased AC signal (Fig. 2a). The higher amplitudes of this AC signal were required to drive the enrichment of the 500-nm particles at the outlet of the post array (as it was done with the 1-μm particles). The enrichment of the 500-nm particles is demonstrated by the increase in magnitude of the red CTF signal vs. time (Fig. 2d). The red CTF signal of the 500-nm particles reached a maximum limit after 32 seconds of starting the elution (after 182 seconds overall). In the final step of this separation, the 500-nm particles were released as a peak of enriched particles by applying the DC voltage for 30 seconds (Fig. 2e). This second released peak also has a well-defined Gaussian shape with relatively small variance. These results demonstrate that the technique has the potential to elute large particles first and can be adequately tuned to perform the selective enrichment and separation of particle mixtures.

Enrichment and separation of the yeast cells and 2-μm particles mixture

A mixture of yeast (average diameter of 6.3 μm) and 2-μm polystyrene particles was studied. First, yeast cells were enriched and eluted using a squared electrical signal with $+V_p$ = 200 V, $-V_p$ = -400 V and ω = 1.2 Hz (Fig. 3a). The subsequent release of the yeast cells as a peak of enriched cells was done using a DC signal of 40 V.

In the initial step of the experiment, yeast cells quickly concentrated at the outlet of the post array. This can be observed by the steep increase in **green** CTF vs. time (Fig. 3b). The 2-μm particles were not enriched, as observed in the **red** CTF signal (Fig. 3b). However, after 25 seconds the concentration of 2-μm particles slightly increases and the enrichment of yeast cells slows down. The size variation in the band of trapped yeast cells was small after 40 seconds, and the electric signal was switched to the DC voltage in order to release the yeast cells as a peak of concentrated cells. Such peak is still unimodal, but it also shows a positive skewed distribution (Fig. 3c). This is caused by a large velocity distribution during the application of the DC voltage, since the once compact band of trapped yeast cells was quickly dispersed into several clusters.

The viability of the yeast cells was quantified to support three important aspects of the technique. First, since the technique allows for a much quicker elution of cells, it is gentler with cells. The elution of yeast cells was achieved after only 40 seconds of applying the DC-biased AC signal, where 85±9% of the yeast cells remained viable. In a previous study by our group with similar conditions [19], yeast cells were trapped in a device with diamond-shaped posts using a potential of 300 V. In that study, only 67±2% of the yeast cells remained viable after 60 seconds of the DC electric potential application [19]. An estimate of 73% of the yeast cells would be expected to remain viable after 40 seconds, assuming an exponential decay in cell viability. Once the yeast cells arrive to the outlet of the post array, they only experience DEP forces during the negative half-cycle of the electric potential signal. In addition, this force comes from the weaker long side of the post. Cells just migrate along with the fluid during the positive half-cycle of the signal. Thus, cells only experience DEP forces during half of the time period of the electric signal, which is the reason behind the gentler manipulation of cells when using asymmetric posts and DC-biased AC potentials.

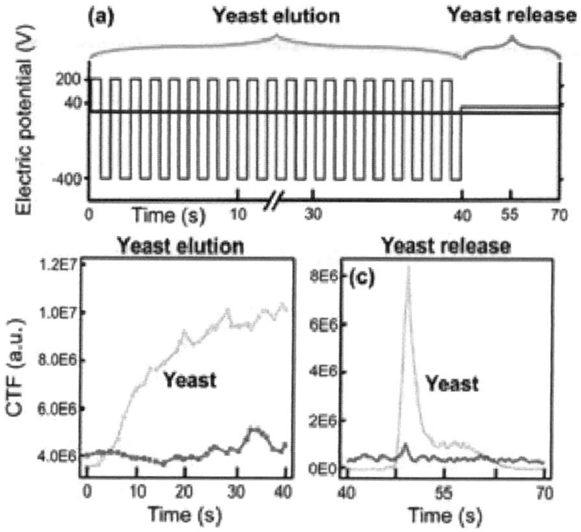

Figure 3: (a) DC-biased AC and DC electric potential signals applied during the elution and release of the yeast cells from a mixture with 2-μm polystyrene particles. **(b-c)** CTF of the yeast cells and 2-μm particles with respect to time during the **(b)** yeast elution and **(c)** yeast release.

Moreover, the technique also reverses the elution order, and cells are exposed to the electric fields for a shorter time and are eluted quickly (40 seconds). The separation of yeast cells from a mixture with 1-μm and 2-μm polystyrene particles was previously reported using diamond-shaped insulating structures [17]. A DC-biased AC signal ($+V_p$ = 900 V, $-V_p$ = -600 V and ω = 20 Hz, V_{RMS} = 765 V) was applied for 60 seconds to elute the 1-μm particles and was followed by a DC voltage of 200 V during 45 seconds to elute the 2-μm particles. The yeast cells were released last, after 105 seconds of operation. Only 32% of the yeast cells would be expected to be viable after such longer electrical treatment and high electric field intensities. Since yeast cells are the larger particles, they would be eluted first using the proposed approach (after 40 seconds) with an average of 85% of viable yeast cells.

Conclusion

A novel dielectrophoretic approach that has the characteristic of allowing quick enrichment and elution of larger, and perhaps more fragile, particles in a sample was presented. This is especially important when working with mixtures that contain eukaryotic cells, which can be easily damaged [26,20,27]. Using asymmetric posts coupled with DC-biased AC potentials creates dissimilar DEP forces that can selectively concentrate the larger particles in the mixture at the outlet of the post array. The approach also drives the smaller particles toward the inlet. The enriched larger particles can be quickly released as a peak of concentrated particles. A key aspect of this separation is the use of the negative DC offset; this offset causes an instantaneous displacement of the DEP barrier generated by the sharp side of the posts in the upstream direction. The same negative DC offset also promotes a net displacement towards the inlet for the smaller particles. The separation of a mixture with yeast cells as the larger particles showed a successful separation of yeast cells in under 40 seconds, while retaining a high cell viability of 85%. These results demonstrate that the use of asymmetric insulating posts combined with low frequency DC-biased AC potentials may be a preferable iDEP separation technique when the viability of larger cells in biological and clinical samples a priority.

Acknowledgements

The authors would like to acknowledge the financial support provided by the National Science Foundation (Award CBET-1336160).

References

[1] Jones, T. B., Electromechanics of Particles, Cambridge University Press, USA, 1995.

[2] Srivastava, S., Gencoglu, A., Minerick, A., *Anal. Bioanal. Chem.* 2010, *399,* 301-321.

[3] Weiss, N. G., Jones, P. V., Mahanti, P., Chen, K. P., Taylor, T. J., Hayes, M. A., *Electrophoresis* 2011, *32,* 2292-2297.

[4] Pysher, M. D., Hayes, M. A., *Anal. Chem.* 2007, *79,* 4552-4557.

[5] Lapizco-Encinas, B. H., Simmons, B. A., Cummings, E. B., Fintschenko, Y., *Electrophoresis* 2004, *25,* 1695-1704.

[6] Moncada-Hernández, H., Lapizco-Encinas, B. H., *Anal. Bioanal. Chem.* 2010, *396,* 1805-1816.

[7] Lapizco-Encinas, B. H., Simmons, B. A., Cummings, E. B., Fintschenko, Y., *Anal. Chem.* 2004, *76,* 1571-1579.

[8] Kang, Y., Li, D., Kalams, S., Eid, J., *Biomed. Microdev.* 2008, *10,* 243-249.

[9] Abdallah, B. G., Chao, T.-C., Kupitz, C., Fromme, P., Ros, A., *ACS Nano* 2013, *7,* 9129-9137.

[10] Nakano, A., Camacho-Alanis, F., Chao, T.-C., Ros, A., *Biomicrofluidics* 2012, *6,* 034108-034113.

[11] Nakano, A., Chao, T.-C., Camacho-Alanis, F., Ros, A., *Electrophoresis* 2011, *32,* 2314-2322.

[12] Gallo-Villanueva, R. C., Rodríguez-López, C. E., Díaz-de-la-Garza, R. I., Reyes-Betanzo, C., Lapizco-Encinas, B. H., *Electrophoresis* 2009, *30,* 4195-4205.

[13] Camacho-Alanis, F., Gan, L., Ros, A., *Sens. Actuator B-Chem.* 2012, *173,* 668-675.

[14] Luo, J., Abdallah, B. G., Wolken, G. G., Arriaga, E. A., Ros, A., *Biomicrofluidics* 2014, *8,* 021801.

[15] Lapizco-Encinas, B. H., Davalos, R., Simmons, B. A., Cummings, E. B., Fintschenko, Y., *J. Microbiol. Methods* 2005, *62,* 317-326.

[16] Masuda, T., Maruyama, H., Honda, A., Arai, F., *PLoS ONE* 2014, *9,* e94083.

[17] Gencoglu, A., Olney, D., LaLonde, A., Koppula, K. S., Lapizco-Encinas, B. H., *Electrophoresis* 2014, *35,* 363-373.

[18] Saucedo-Espinosa, M. A., Lapizco-Encinas, B. H., *Electrophoresis* 2015, *36,* 1086-1097.

[19] LaLonde, A., Romero-Creel, M. F., Lapizco-Encinas, B. H., *Electrophoresis* 2014, *36,* 1479-1484.
[20] Teissié, J., Rols, M. P., *Biophys. J.* 1993, *65,* 409-413.
[21] Du, F., Ciaciuch, P., Bohlen, S., Wang, Y., Baune, M., Thöming, J., *J. Membr. Sci.* 2013, *448,* 256-261.
[22] LaLonde, A., Gencoglu, A., Romero-Creel, M. F., Koppula, K. S., Lapizco-Encinas, B. H., *J. Chromatogr. A* 2014, *1344,* 99-108.
[23] Schindelin, J., Arganda-Carreras, I., Frise, E., Kaynig, V., Longair, M., Pietzsch, T., *et al.*, *Nat Meth* 2012, *9,* 676-682.
[24] Duffy, D. C., McDonald, J. C., Schueller, O. J. A., Whitesides, G. M., *Anal. Chem.* 1998, *70,* 4974-4984.
[25] Saucedo-Espinosa, M. A., Rauch, M., Lapizco-Encinas, B. H., in *AICHE Annual Meeting 2014*, AICHE, Atlanta GA, USA, November 2014, 2014.
[26] Potter, H., Heller, R., *Transfection by Electroporation*, In F.M. Ausubel (Editor), *Current Protocols in Molecular Biology*, John Wiley and Sons, Inc., 2003.
[27] Gallo-Villanueva, R. C., Sano, M. B., Lapizco-Encinas, B. H., Davalos, R., *Electrophoresis* 2014, *35,* 352-361.

428484 Rapid Culture-Based Detection of Mycobacteria Using Electrical Impedance Spectroscopy (EIS) Measurements

Abstract:

Tuberculosis (TB) is a bacterial infection which commonly affects the pulmonary system. According to a WHO's Global TB report 2014, in 2013, about 9 million people were detected with TB and about 1.5 million people died. However, if detected early, TB is treatable. Treating TB is possible if the TB *mycobacterium* in the clinical samples can be detected rapidly.

Till date, culture based detection of TB *mycobacteria* is the most common mode of detection and is the current gold standard. This method is cheap and reliable. However, it is a time consuming procedure and may take anywhere from 7-56 days as pathogens like *Mycobacterium* species are slow growing organisms (having a doubling time of ~24 hours vs ~20mins for *Escherichia coli*). Here we show the feasibility of using multi-frequency EIS based to detect TB *mycobacterium* in a significantly shorter amount of time.

To conduct our *in vitro* time-to-detection (TTD) study, *Mycobacterium bovis* (an analog) was used instead of *Mycobacterium tuberculosis*. *M. bovis* was cultured in modified Middlebrook 7H9 mycobacteria media at 37C. It was also grown in Lowenstein-Jensen plates. For our study, three different concentrations of the cultured *M. bovis* were seeded in the growth media: low (having 2-5 CFU/ml), medium (having 100-200 CFU/ml and high (having ~3000 CFU/ml). In addition, a control sample (with no bacteria) was also studied. One set of samples (control, low, medium and high concentrations) were also sent to the University of Missouri Hospital's microbiology laboratory for detection using the current state of the art automated culture-based system; viz. the BACTEC MGIT^{TM} 960 from Becton Dickinson. The various concentration samples and control samples were all incubated under similar conditions of 37C and periodically small amounts of aliquots were withdrawn from it and put on a 3D printed microfluidic channel to analyze the bulk capacitance of the suspension using multi-frequency EIS measurements. The microfluidic channel is designed to have a thin capillary channel with two gold electrodes inserted in it such that the distance between the electrodes is 1 cm. During EIS measurement, AC voltage is applied to the electrodes and the bulk capacitance of the suspension resulting from the charges that accumulated in the bacterial cells is recorded at multiple frequencies ranging from 1 KHz to 100MHz.

For the control samples, it is observed that subsequent bulk capacitance readings remain statistically unchanged from the baseline reading. On the other hand, bulk capacitance recorded for samples containing *Mycobacteria*, show gradual increase in the value, and become statistically significantly different from the baseline reading when the concentration of the suspension reaches 10^3-10^4 CFU/ml. Plate counts of the aliquots withdrawn are done to get the concentration values. Using our method, the TTD for high concentration samples were about 60-70 hours (< 3 days), medium concentration samples were about 84-108 hours (< 5 days) and for low concentration samples were about 132-156 hours (< 7 days). In contrast, the TTD obtained for the same samples from the hospitals were about 5-6 days for high concentrations. TTD for the samples with medium and low concentrations is greater than 40 days (we are still awaiting results).

Hence we have demonstrated a potential technique which has much lower TTD than the current automated systems. Less TTD implies faster clinical response resulting in improved patient response.

Counterflow Isotachophoresis of Pyronin Y in a PMMA Microchip

Lang Qin and Cornelius F. Ivory

Gene and Linda Voiland School of Chemical Engineering and Bioengineering,
Washington State University, Pullman, WA, 99164-6515

Introduction

The use of nuclear fission has led to the global spread of both natural and artificial f-elements in the environment at ultra-trace levels. The separation of f-elements is extremely difficult because of their chemical and physical similarities.[1]

Isotachophoresis (ITP), an electrophoretic separation technique, is an effective method to separate and detect trace f-elements qualitatively and quantitatively based on their different *effective* electrophoretic mobilities as shown in Fig. 1.

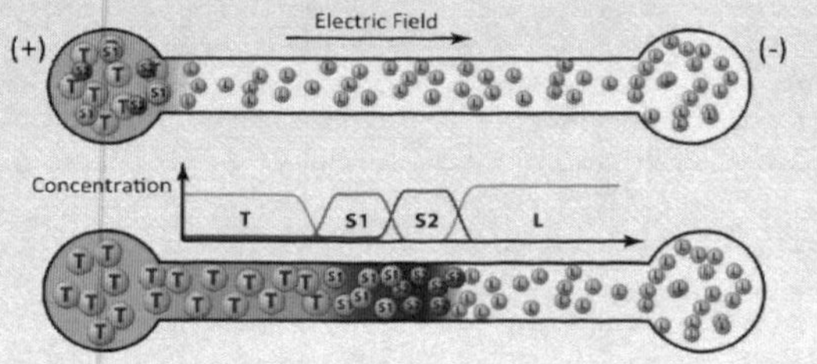

Figure 1. Under an applied electric field, the sample cations (S1 and S2) overspeed the slow trailing electrolyte (T), and self-segregate into discrete zones behind the fast leading electrolyte (L). Adapted from Santiago et al. [2]

The poly(methyl methacrylate) (PMMA) microfluidic platform is portable, inexpensive and fast.[3]

In theory, hydrodynamic counterflow is able to improve the sensitivity of plateau mode ITP by increasing the effective separation length of the channel. Pyronin Y, a cationic dye, can be used as both a simulant and a marker for *f*-element ITP .

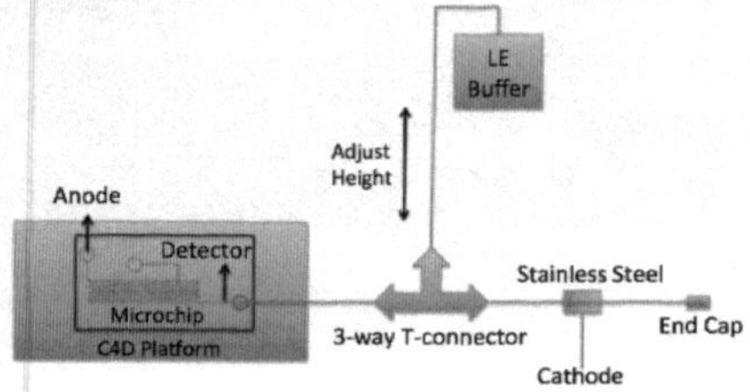

Figure 2. Experiment setup diagram for Counterflow ITP in microfluidic chip system. The pyronin Y signal can be detected with capacitively-coupled contactless conductivity detection (C4D). The velocity of hydrodynamic counterflow can be adjusted by the height of leading electrolyte source.

Hypothesis

We hypothesize that during ITP, a hydrodynamic counterflow generated by a reservoir height difference (see Fig. 2) will produce *stationary* ITP zones of pyronin Y, which will allow us to use shorter channels and will provide more time for the ITP zones to form, thus leading to better resolution between the zones. A 2D model of ITP in a serpentine channel simulates the experimental results using COMSOL Multiphysics v5.1 and shows the impact of counterflow on microchip ITP.

Microchip Fabrication

The PMMA microchip is fabricated by photoresist etching, hot embossing and thermally-activated solvent bonding procedure as shown in Fig. 3.

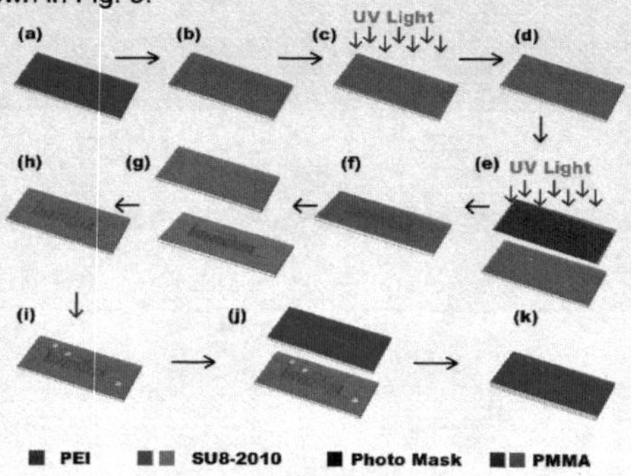

Figure 3. PEI-substrate-based PMMA microchip fabrication process: *(a)-(c) generate sacrificial layer on PEI substrate to increase adhesive force on photoresist, (d)-(f) develop positive microchannel pattern on the PEI substrate, (g)-(h) imprint PMMA slide with PEI substrate, (i) drill reservoirs, (j)-(k) thermal bonding imprinted PMMA slide with blank PMMA slide.*

Results

- Pyronin Y can be concentrated into a pure zone within 5 min at an applied voltage of 5.6 kV.
- Hydrodynamic counterflow is able to produce a stationary ITP zone and push it to any position in the microchannel.
- Our 2D simulation is consistent with our experimental observations.

Results

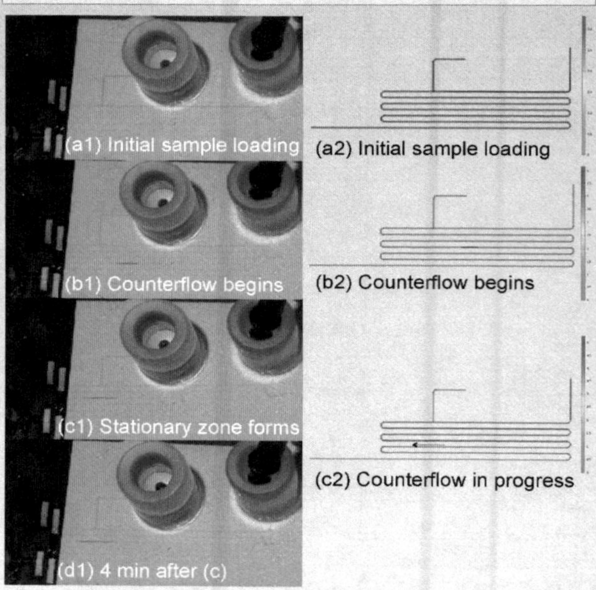

Figure 4. Experiment and simulation results for counterflow ITP. Initial pyronin Y sample loading: t=0 for both (a1) experiment and (a2) model; counterflow starts: (b1) t=76s for model and (b2) t=115s for experiment; stationary zone forms: (c1) t=370s for experiment; counterflow in progress: (c2) t=358s for model; additional dispersion: (d1) t=610s for experiment. The legend is for the concentration of pyronin Y in mM.

References

[1] Janos, P., *Electrophoresis*, 2003, 24,1982-1992.
[2] High-Sensitivity Isotachophoretic Stacking CE Devices, accessed 2015 Nov 05, http://microfluidics.stanford.edu/Projects/Current/HighSensITP.html
[3] Cong, Y,. et al., *Electrophoresis*, 2014, 35, 646–653.

Acknowledgement

This work is supported in part by the Defense Threat Reduction Agency under Grant No. HDTRA 1-14-10069.

Effective Diffusivity and Optimal Time of Separation Under Non-Newtonian Behavior of Fluid Flow Fractionation

Stephen Dueck,[1] Steffano Oyanader,[1] and Mario Oyanader[1]

1. California Baptist University, Riverside, California

Abstract

The separation of pharmaceuticals and metabolites for treatment and diagnosis from bio-samples is carried on cylindrical devices. The Non-Newtonian behavior of the carrier fluid in this case is often neglected to reduce the complexity of the analysis and mathematical derivation. Following this approach, in electrical field flow fractionation (EFFF) several parameters have been investigated to optimize molecular separation [Pascal, Oyanader, and Arce, 2008 and 2010]. The main focus of this contribution is to extend the knowledge and understanding of the carrier fluid effect on the effective diffusivity and optimal time of separation for a more realistic case as that of Non-Newtonian fluids. Analytical expression for the most important parameters are reported and have been obtained using the spatial area averaging methodology. The analysis has been illustrated by selecting parameter values that represent a number of potential useful applications. A set of five parameters has been combined to obtain the best operating conditions for optimal separation of solutes. Effective transport parameters are used to study the role of the power law "n" parameter and of the magnitude of the applied orthogonal electrical field on the values of the optimal time of separation of species on Non-Newtonian (pseudo-plastic and dilatants) applications.

Process Sketch and Diagrams

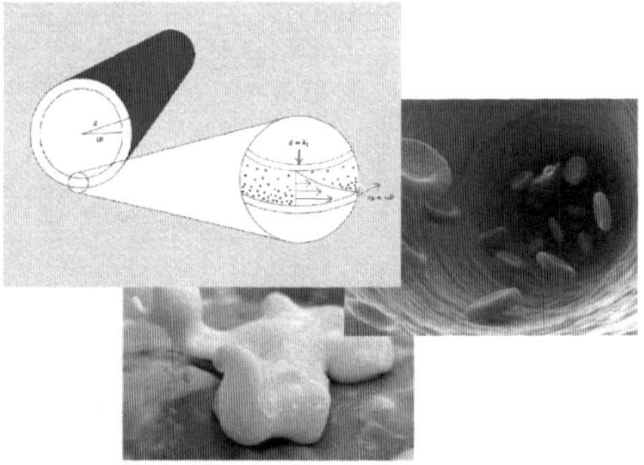

Equations

Carrier Velocity Governing Equation

$$\frac{\partial}{\partial r}\left(r^2 T_{r\theta}\right) = 0$$

Shear Stress Power Law Model

$$T_{r\theta} = -m\left(r\frac{\partial}{\partial r}\left(\frac{v_\theta}{r}\right)\right)^n$$

Boundary Condition

$$v_\theta = 0 @ r = kR$$
$$v_\theta = \omega R @ r = R$$

Electrical Field Governing Equation

$$\frac{1}{r}\frac{\partial}{\partial r}\left(r\frac{\partial \phi}{\partial r}\right) = 0$$

Boundary Conditions

$$\phi = 0 @ r = kR$$
$$\phi = K_1 @ r = R$$

Species Continuity Governing Equation

$$\frac{\partial C}{\partial t} + \frac{v_\theta}{r}\frac{\partial}{\partial \theta} = D\left[\frac{1}{r}\frac{\partial}{\partial r}\left(r\frac{\partial C}{\partial r}\right) + \frac{1}{r^2}\frac{\partial^2 C}{\partial \theta^2}\right] + \mu^* \frac{1}{r}\frac{\partial}{\partial r}\left(rC\frac{\partial \phi}{\partial r}\right)$$

Boundary Conditions

$$D\frac{\partial C}{\partial r} + \mu^*\left(C * \frac{\partial \phi}{\partial r}\right) = 0 @ r = kR$$
$$D\frac{\partial C}{\partial r} + \mu^*\left(C * \frac{\partial \phi}{\partial r}\right) = 0 @ r = R$$

Effective Dispersion Coefficient

$$\Theta = \frac{A_0}{2(2-\Omega)}\left\{\frac{(k^4 - k^4)}{(k^2 - k^4)}\frac{(A-2)}{(A-4)}A_0 - \frac{1}{2}(k^2+1)A_1 f\right\} - \left\{\frac{A_0}{2\Omega^*} + \frac{A_1}{(2-\Omega^*)\Omega^*}\right\}\{A_0 + A_1 f\} - \frac{A_1}{(2-A)(2-A-\Omega^*)}\left\{\frac{(k^{2-A}+1)A_0}{2} - \frac{2(k^{4-A}-1)}{(k^2-1)(A-4)}A_1 f\right\}$$

$$A_0 = \frac{2k^2 - 2k^{\frac{2}{\pi}}}{\frac{2}{\pi} - 2 - \frac{2}{\pi}k^2 + 2k^{\frac{2}{\pi}}} \qquad f = \frac{\left(k^{2+\frac{\Omega}{m(k)}} - k^{\frac{2}{\pi}}\right)\left(2 + \frac{\Omega}{m(k)}\right)}{k^{\frac{2}{\pi}}\left(k^{2+\frac{\Omega}{m(k)}} - 1\right)\left(2 + \frac{\Omega}{m(k)} - \frac{2}{\pi}\right)} \qquad A_1 = \frac{\left(\frac{2}{\pi} - 2\right)(k^2 - 1)k^{\frac{2}{\pi}}}{\frac{2}{\pi} - 2 - \frac{2}{\pi}k^2 + 2k^{\frac{2}{\pi}}}$$

$$\Omega \equiv \frac{\mu * K_1}{D}$$

Optimal Separation Time

$$\tau_{op} = \left[\frac{\sqrt{\frac{D_{effA}}{D_A}} + \sqrt{\frac{D_{effB}}{D_B}}}{\hat{v}_{effA} - \hat{v}_{effB}}\right]^2$$

Where:

$$\frac{D_{eff}}{D} = 1 - \theta \cdot Pe^2$$
$$\hat{v}_{eff} = 1 + A_0 + A_1 * f$$

Illustrative Cases

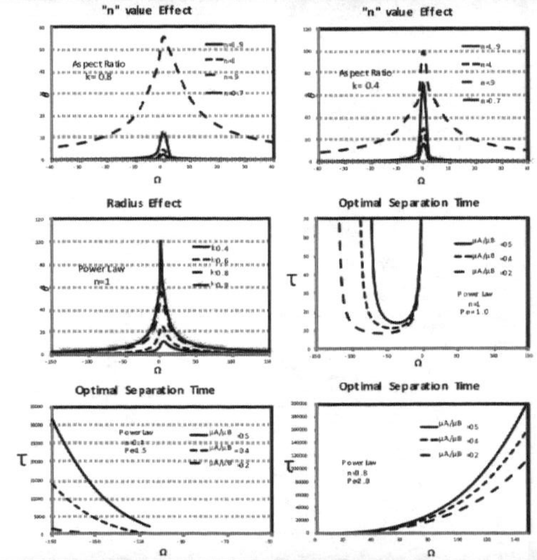

Conclusions

- Effective Dispersion Coefficient values decrease as the Aspect Ratio parameter (k) increases.
- Non Newtonian behavior has a significant influence on the magnitude of Effective Dispersion Coefficient values and therefore make separation impossible given the large magnitude of the Optimal Separation Time.

References

Pascal, J., Oyanader, M. A. and Arce, P., Effect of Wall Velocities on the Determination of Optimal Separation Times in Electrical Field Flow Fractionation, The Canadian Journal of Chemical Engineering, vol. 88, pp. 384-391, 2010.

Oyanader, M. A. and Arce, P., Role of Geometrical Dimensions In Electrophoresis Applications With Orthogonal Fields, Electrophoresis, vol. 26, pp. 2857-2866, 2005.

cbu Gordon and Jill Bourns College of Engineering

AIChE The Global Home of Chemical Engineers

Effect of Particle Concentration in the Estimation of Dielectrophoretic Force in Arrays of Insulators

Mario A. Saucedo-Espinosa and Blanca H. Lapizco-Encinas*

Microscale Separations Laboratory, Rochester Institute of Technology, NY, United States

Introduction

Microfluidics, the field of science that studies the manipulation of small volumes of liquid, is a constantly expanding field. Numerous microfluidic devices have been used to date in several areas, such as the analysis of particles and cells in clinical and environmental analyses [1,2]. Electric fields driven techniques is one of the most important microfluidic areas, due to their ease of applications and robustness [3]. Among the electric field driven techniques, dielectrophoresis (DEP) has become a popular electrokinetic transport mechanism given that provides means for simultaneous particle sorting and enrichment of particles and cells [4]. Dielectrophoresis exploits particle polarization effects, which occur when particles are exposed to a non-uniform electric field, the response can be positive or negative DEP behavior. Each behavior refers to particles attracted or repelled from the regions of higher electric field gradient [5].

Particles being manipulated in insulator-based DEP (iDEP) devices also experience electrophoresis (EP) and electroosmotic (EO) flow effects. Since EP and EO flow are linear functions of the electric field, their contribution can be considered in a single electrokinetic (EK) transport term. At low applied voltages, usually EK dominates particle movement and particles flow along with the fluid through the channel. In contrary, at higher applied voltages, particles are trapped or retained when DEP overcomes EK [6].

Numerous studies have reported the capabilities of DEP systems in the detection, sorting, enrichment and isolation of biological particles, such as proteins [7,8], DNA

*Email: bhlbme@rit.edu

[9,10], virus [9] and bacteria [11-15]. Further, DEP has also been used to manipulate non-biological particles, like colloids [16-18], carbon nanotubes [19] and polystyrene particles [20-22]. Even if there are many reports on the experimental trapping capacity of DEP systems, most particle trapping analysis haven been qualitative [23]. Uncertainty in the physical properties (*e.g.*, permittivity, size, deformability and local field gradients) [24], particle interactions [23], simplifications in particle shape [25] and in the use of the dipole approximation [26] pose a challenging derivation of an accurate estimation of the DEP force.

Despite the significant contributions made by many studies, there is a need to understand the phenomena that gives rise to the need of the correction factor. This contribution presents a systematical study of the correction factor as a function of particle concentration to elucidate possible mechanisms that give rise to a significant deviation from theory; the correction factor was estimated by matching the experimental DEP force in a tampered microchannel [24] with numerical simulations.

Theoretical background

Electrokinetics and dielectrophoresis

Particle movement in iDEP devices is influenced by EP and EO flow. Electrokinetics (EK) is defined as the superposition of EP and EO flow [27]:

$$\vec{v}_{ek} = \mu_{ek}\vec{E} = \left(\mu_{eo} + \mu_{ep}\right)\vec{E} = -\frac{\varepsilon_m\left(\zeta_w - \zeta_p\right)}{\eta}\vec{E}, \quad (1)$$

where μ_{ek}, μ_{eo} and μ_{ep} are the EK, EO and EP mobilities, respectively; ε_m and η are the real permittivity and viscosity of the suspending medium, ζ_w and ζ_p are the zeta potentials of the wall and particle, respectively, and E denotes the electric field.

The influence of DEP in particles results in their migration motion towards the regions with higher (positive DEP) or lower (negative DEP) electric field gradient. The DEP velocity $\vec{v}_{dep}$ of a spherical particle of radius r_p is defined as [5,27,28]:

$$\vec{v}_{dep} = \mu_{dep}\nabla E^2 = \frac{r_p^2 \varepsilon_m}{3\eta}\mathrm{Re}[f_{CM}]\nabla E^2, \quad (2)$$

where μ_{dep} is the DEP mobility and $\mathrm{Re}[f_{CM}]$ is the real part of the Clausius-Mossotti (CM) factor.

Particle velocity at the centerline

Under the influence of a DC electric field, the net particle velocity of a spherical particle along the longitudinal direction, in the absence of particle-particle interactions, is [24]:

$$v_{p,x} = \left(\mu_{ek} + \mu_{dep}\frac{\delta E}{\delta s}\right)E\cos\theta - \frac{\mu_{dep}E^2}{R}\sin\theta, \quad (3)$$

where s is the arc length along the field line, θ is the angle between the tangent of the field line and the longitudinal axis, and R is the radius of curvature of the field line. If the net particle velocity is analyzed in the centerline of a symmetrical channel, the field lines are parallel to the longitudinal axis, so that $\theta = 0$, and the net particle velocity simplifies to [24]:

$$v_{p,x} = \left(\mu_{ek} + \mu_{dep}\frac{\delta E}{\delta x}\right)E. \quad (4)$$

Particle net velocity in a tampered channel

Weiss *et al.* [24] proposed the use of a channel with three well-defined sections in order to adequately estimate μ_{dep} (Fig. 1). The channel consists of a wide segment with uniform area (first region in Fig. 1), followed by a constricting tapered segment (second region in Fig. 1) and a final narrow segment with uniform area (third region in Fig. 1). Given that regions 1 and 3 have a uniform area, they can be used to estimate μ_{ek}. The second region, which is tapered, was designed to create a linearly increasing electric field (Fig. 1) [24].

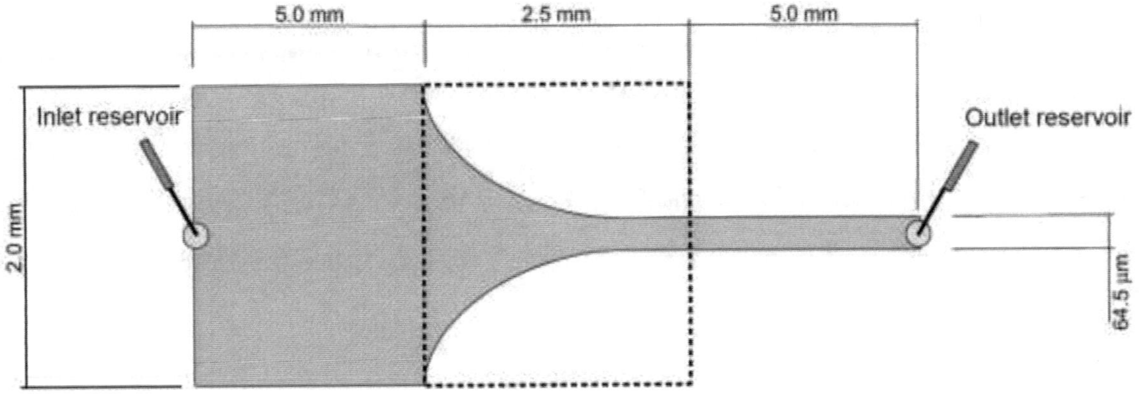

Figure 1. Tempered channel employed in this study and proposed by Weiss et al. [24].

In the tapered region, the channel height (h) is defined by the photoresist thickness and the channel cross-sectional area (A) varies inversely with x [24]:

$$A = h \cdot w_2(x) = h\frac{w_1}{(1+kx)}, \quad 0 \le x \le 2.5\,mm, \tag{5}$$

where w_1 and w_2 are the channel widths of region 1 and region 2, respectively, and k is the rate at which the channel tapers. The taper region starts at $x = 0$ and ends at $x = 2.5$ mm. After considering the charge density, the net particle velocity along the centerline can be defined as [24]:

$$v_{p,x} = (\gamma^2 k^2 \mu_{dep} + \gamma k \mu_{ek})x + (\gamma^2 k \mu_{dep} + \gamma k \mu_{ek}), \tag{6}$$

In equation (6), γ represents the electric field in region 1. Hence, equation (6) can be used to estimate μ_{dep} by analyzing the slope of the net velocity profile along the centerline.

Experimental section

Microparticles

Fluorescent red 5-µm (ex/em 580/605 nm) polystyrene particles from Invitrogen (Carlsbad, CA) were employed for experimentation. The stock suspensions of the microspheres where diluted in 0.2 mM K_2HPO_4 (pH = 7.21, conductivity = 59.3 µS/cm) to final concentrations as described in the results and discussion sections.

Microdevice description

Microchannels were made from polydimethylsiloxane (PDMS, Dow Corning, Midland, MI) and fabricated using standard soft-lithography techniques [29]. All the interior walls were made of PDMS and have the same zeta potential. For the experiments described here w_1 = 5 mm, k = 12.93 mm^{-1}, h = 0.04 mm, γ = 5 V/mm and the taper ended at x = 7.5 mm.

Equipment

A high voltage sequencer (Model HVS6000D; LabSmith, Livermore, CA) was used to apply DC electric potentials by means of platinum wire electrodes. The voltage sequencer was manipulated with the software *Sequence* provided by the manufacturer. The experiments were carried out using a ZEISS Axiovert 40 CFL inverted microscope (Carl Zeiss Microscopy, Thornwood, NY). An Infinity 2 camera (Luminera, Ottawa, Canada) was used to capture microscopy images and videos. The post-processing, including the velocity determination using the streak velocimetry technique [24], was performed using Fiji image processing package [30].

Results and discussion

Particle concentration was varied between 7.59×10^5 and 6.99×10^7 particles/mL. Eight intermediate concentrations, equally spaced, were studied. Figure 2 shows the measurement of the particle velocity with respect to the position of the particles for the two extreme concentrations. As observed, particle velocity in both experiments follows a straight line with respect to particle position. However, data is more spread at the higher concentration. This broader distribution, attributed to noise in the measurements where the trajectories of two particles are indistinguishable from each other as particle concentration increases, results in a poorer evaluation of the DEP underlying DEP mobility.

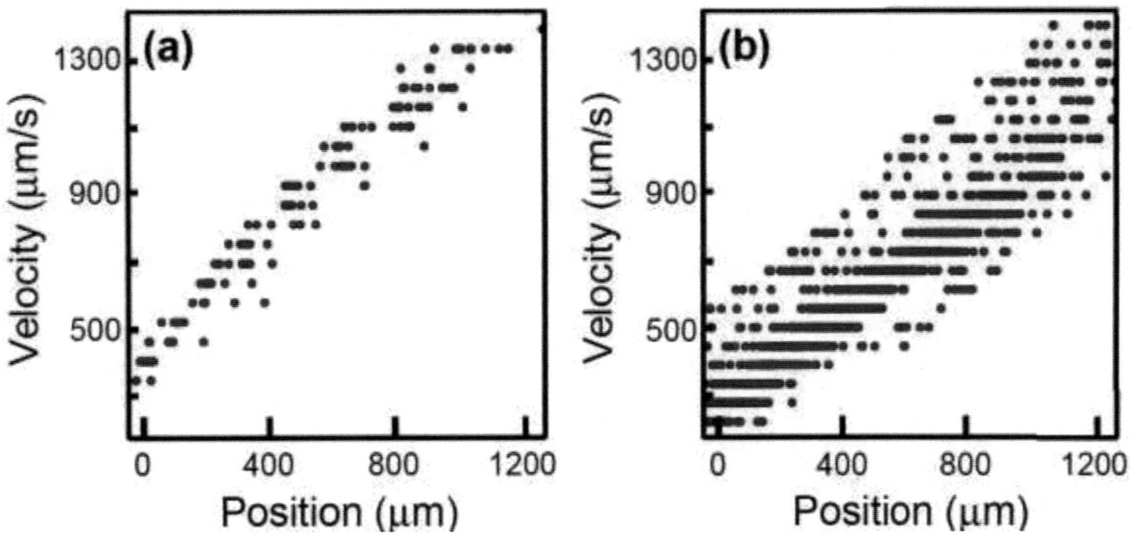

Figure 2. Measured particle velocity as a function of position inside the tampered channel for a particle concentration of (a) 7.59E5 and (b) 6.99E7 beads/mL

Figure 3 shows the DEP mobility estimated using equation (6). The calculated DEP mobilities were normalized with respect to the DEP mobility at the lowest concentration. As expected, particle concentration has an effect on the DEP mobility, where it increases along with particle concentration. Considering the negative DEP behavior of the polystyrene particle employed in this study, this implies that the more particle in the system, the easier it is to trap them, as has been reported in several studies. However, this increase in the estimated DEP mobility is not able to completely explain the two order of magnitude difference reported by Weiss et al. [24].

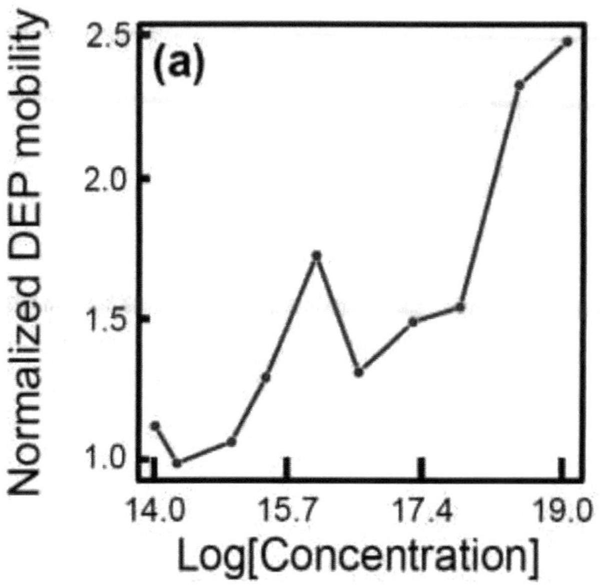

Figure 3. DEP mobility as estimated by fitting particle velocity in the tampered channel as a function of the logarithm of particle concentration. The DEP mobilities are normalized with respect to the lowest concentration.

Conclusion

Based on our preliminary results, particle concentration has an effect on the deviation between the experimental DEP mobility and the theoretical one. This, in turn, affects the estimation of the correction factor. The contribution of mutual particle interaction was shown to increase the DEP mobility up to a factor of 2.5 times. However, the noise in the measured velocity increases with particle concentration, so that further experimentation is needed in order to provide better estimates. By using further experimental points, the underlying function that gives rise to the correction factor is expected to be assessed.

References

[1] Martinez-Duarte, R., *Electrophoresis* 2012, *33,* 3110-3132.

[2] Khoshmanesh, K., Nahavandi, S., Baratchi, S., Mitchell, A., Kalantar-zadeh, K., *Biosensors and Bioelectronics* 2011, *26,* 1800-1814.

[3] Çetin, B., Li, D., *Electrophoresis* 2011, *32,* 2410-2427.

[4] Moncada-Hernández, H., Lapizco-Encinas, B. H., *Analytical and Bioanalytical Chemistry* 2010, *396,* 1805-1816.

[5] Jones, T. B., Electromechanics of Particles, Cambridge University Press, USA, 1995.

[6] Cummings, E. B., *IEEE Engineering in Medicine and Biology Magazine* 2003, *22,* 75-84.

[7] Abdallah, B. G., Chao, T.-C., Kupitz, C., Fromme, P., Ros, A., *ACS Nano* 2013, *7,* 9129-9137.

[8] Lapizco-Encinas, B. H., Ozuna-Chacón, S., Rito-Palomares, M., *Journal of Chromatography A* 2008, *1206,* 45-51.

[9] Sonnenberg, A., Marciniak, J. Y., McCanna, J., Krishnan, R., Rassenti, L., Kipps, T. J., Heller, M. J., *Electrophoresis* 2013, *34,* 1076-1084.

[10] Martinez-Duarte, R., Camacho-Alanis, F., Renaud, P., Ros, A., *Electrophoresis* 2013, *34,* 1113-1122.

[11] Lapizco-Encinas, B. H., Simmons, B. A., Cummings, E. B., Fintschenko, Y., *Electrophoresis* 2004, *25,* 1695-1704.

[12] Bernabini, C., Holmes, D., Morgan, H., *Lab on a Chip* 2011, *11,* 407-412.

[13] Braff, W. A., Willner, D., Hugenholtz, P., Rabaey, K., Buie, C. R., *PLoS ONE* 2013, *8,* e76751.

[14] Hamada, R., Takayama, H., Shonishi, Y., Mao, L., Nakano, M., Suehiro, J., *Sensors and Actuators B: Chemical* 2013, *181,* 439-445.

[15] Park, S., Zhang, Y., Wang, T.-H., Yang, S., *Lab on a Chip* 2011, *11,* 2893-2900.

[16] McMullan, J. M., Wagner, N. J., *Langmuir* 2012, *28,* 4123-4130.

[17] Yunus, N. A. M., Nili, H., Green, N. G., *Electrophoresis* 2013, *34,* 969-978.

[18] Jain, S., Gupta, S., *Langmuir* 2013.

[19] Duchamp, M., Lee, K., Dwir, B., Seo, J. W., Kapon, E., Forró, L., Magrez, A., *ACS Nano* 2010, *4,* 279-284.

[20] White, C. M., Holland, L. A., Famouri, P., *ELECTROPHORESIS* 2010, *31,* 2664-2671.

[21] Gallo-Villanueva, R., Pérez-González, V. H., Davalos, R., Lapizco-Encinas, B. H., *Electrophoresis* 2011, *32,* 2456–2465.

[22] Gencoglu, A., Olney, D., LaLonde, A., Koppula, K. S., Lapizco-Encinas, B. H., *Electrophoresis* 2014, *35,* 363-373.

[23] Rosenthal, A., Taff, B. M., Voldman, J., *Lab on a Chip* 2006, *6,* 508-515.

[24] Weiss, N. G., Jones, P. V., Mahanti, P., Chen, K. P., Taylor, T. J., Hayes, M. A., *Electrophoresis* 2011, *32,* 2292-2297.

[25] Saucedo-Espinosa, M. A., Lapizco-Encinas, B. H., *Electrophoresis* 2015, *36,* 1086-1097.

[26] Washizu, M., Jones, T. B., *Journal of Electrostatics* 1994, *33,* 187-198.

[27] Probstein, R. F., Physicochemical hydrodynamics: an introduction, John Wiley & Sons, 2005.

[28] Pohl, H. A., Dielectrophoresis, Cambridge University Press, Cambridge, 1978.

[29] Duffy, D. C., McDonald, J. C., Schueller, O. J. A., Whitesides, G. M., *Analytical Chemistry* 1998, *70,* 4974-4984.

[30] Schindelin, J., Arganda-Carreras, I., Frise, E., Kaynig, V., Longair, M., Pietzsch, T., Preibisch, S., Rueden, C., Saalfeld, S., Schmid, B., Tinevez, J.-Y., White, D. J., Hartenstein, V., Eliceiri, K., Tomancak, P., Cardona, A., *Nat Meth* 2012, *9,* 676-682.

Dielectrophoretic Separation of Healthy and Infected Red Blood Cells through Electric Driven Flow: An Electrokinetic Modeling

University of Idaho College of Engineering

Milad Nahavandi[1], Soumya K. Srivastava[1]
[1]Department of Chemical and Materials Engineering
University of Idaho

MESA Lab
Microfluidic Electrokinetic bioSeparations and Analysis Lab

ABSTRACT

A new approach to insulating dielectrophoretic cell sorting using numerical modeling and simulation of electrokinetic bio-particle movements in micro-channel is presented. We report a microfluidic device capable of separating healthy red blood cells (h-RBCs) from RBCs infected with Babesiosis based on the simultaneous application of electrokinetic and dielectrophoretic forces acting on cells. After validation of numerical results with experimental data, a particle sorting model was developed to evaluate the effects of channel geometry, applied voltage and medium pH strength. Numerical investigations were performed based on the suggested model and the theoretical cell trajectory was calculated. Simulation results showed that RBCs with different dielectric properties are influenced by different dielectrophoretic forces while they are continuously pushed by electrophoresis and fluid drag force caused by electroosmotic flow over wall surface, thus directing to different streamlines in microchannel. Also, it was indicated that the pH of medium substantially influenced the zeta potential of RBCs and microchannel surface, which can widely change the electrokinetic movement of cells. Results also showed that a perfect separation can be achieved at pH 5 and applied DC voltage of 15 V.

PROJECT OVERVIEW AND DESCRIPTION

INTRODUCTION

- LOC devices using insulating DEP for manipulation and separation of bio-particles have been noticeably developed over the past decade.
- They provide non-invasive, cheaper, low-energy requirement, portable, and quick identification for accurate disease diagnostics.
- Numerical modeling and simulations using COMSOL Multiphysics V. 5.1 software was used to evaluate the effect of different operating conditions on separation efficiency.
- This device is benefited from negative Dielectrophoretic (n-DEP) force generated by DC applied voltage to manipulate RBCs.

THEORY

DC Dielectrophoretic Force Acting on RBCs

$$F_{DEP} = 2\pi r_p^3 \varepsilon_0 \varepsilon_m f_{CM} \nabla |E|^2 \qquad f_{CM} = \frac{\varepsilon_p - \varepsilon_m}{\varepsilon_p + 2\varepsilon_m}$$

F_{DEP} (Healthy RBCs) = 2.75 nN, F_{DEP} (Infected RBCs) = 1.5 nN , μ_{EO} = 10^{-8} [m²/V.s], μ_{EP} = 1.14*10^{-8} [m²/V.s]

DEVICE DESIGN

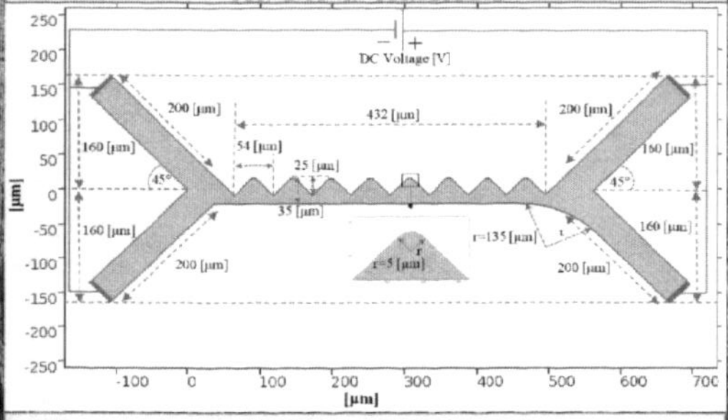

MODEL APPROACH

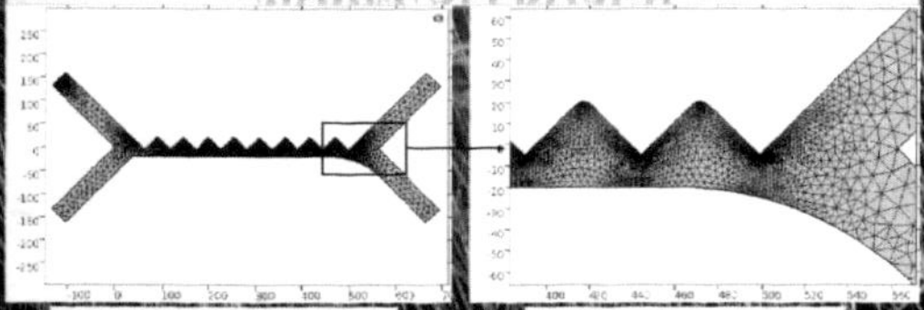

Triangular advanced front mesh

Finer mesh around hurdles

Type of Elements	Elements Parameters	
	Min. Size [μm]	Max. Size [μm]
Bulk Elements	1.1	10
Elements used for hurdles peak	0.2	1

Type of Cells		Diameter [μm]	Relative permittivity, ε_r	Conductivity, σ
h-RBCs		7	60	0.35
Babesiosis	Host cell	7	59	-
	Parasite	1	68	1.2
Suspending medium (dilute blood plasma)		-	75	0.05

RESULTS AND DISCUSSION

Three dependencies investigated: 1) DC Voltage, 2) Medium pH, 3) Hurdle No.
Flow and electric field profiles were observed and cells were traced by particle tracing module.

ELECTRIC FIELD

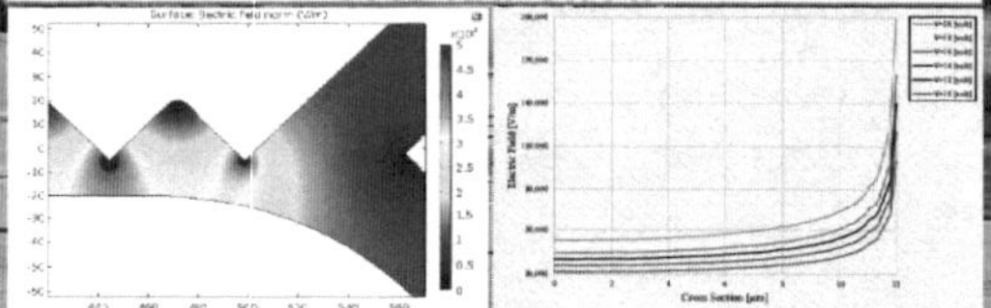

- There is no dielectric loss due to no frequency component in the electric field.
- DC voltage of < 20 [V] eliminates the irreversible damage to the cell membrane.
- Strong field gradient at hurdle peaks facing channel surface.

VELOCITY VARIATION

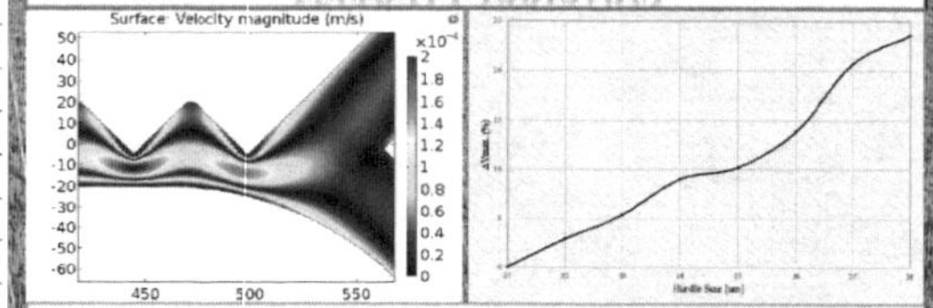

- Electrokinetic flow is generated through the microchannel due to 1) electro-osmotic induced flow over the PDMS surface and 2) flow due to electrophoretic particle movement.
- The electro-osmotic induced flow results from the drag force acting on RBCs in motion experiencing DEP force at the hurdles.
- Electro-osmotic surface flow effect on the cells is only observed at the walls and is reduced as the cells move away from the walls.

PARTICLE TRACING

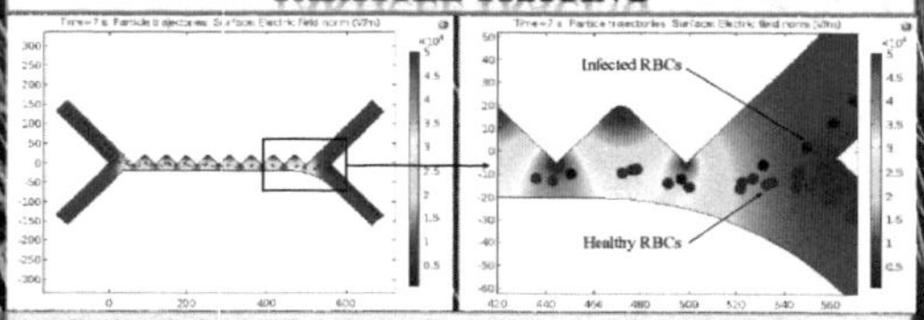

- Healthy and infected RBCs motion is influenced by the combined effects of electro-osmosis and electrophoresis and separated by DEP force when the cells are near the hurdles.
- Healthy RBCs are pushed much further away down-stream compared to the infected RBCs going into upper channel due to more n-DEP they are influenced.
- Complete separation is achieved by applying low DC of ~15 V.

EFFECT OF HURDLES NUMBER

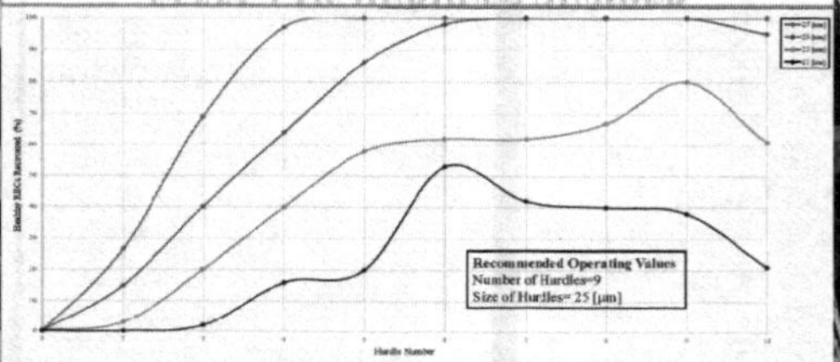

EFFECT OF ELECTRIC VOLTAGE

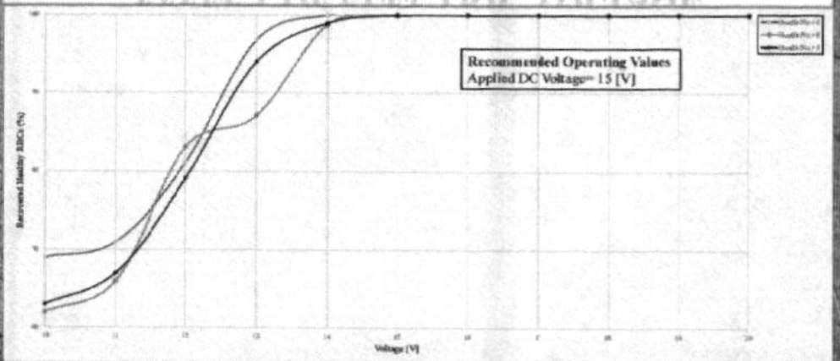

EFFECT OF MEDIUM pH

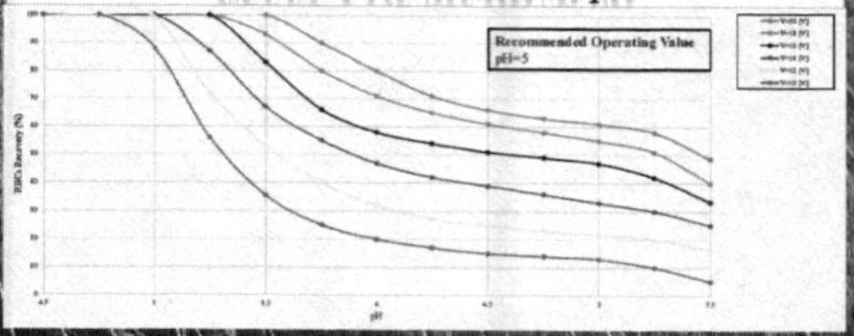

CONCLUSION

- Applying DC voltage on a smaller device platform results in strong electric field around hurdles.
- Electrokinetic flow is generated over the microchannel surface that eventually leads to bulk flow resulting in cell motion.
- Sorting of cells is under effect of hurdles No. with the optimum # 9, and having lesser distance between the hurdle and the surface increases sensitivity and efficiency by 70%.
- Perfect separation of healthy and infected RBCs is obtained at an optimum applied DC of 15 [V] over 800 μm length of device platform.
- The change in suspending medium pH substantially affects the electrokinetic velocity of cell suspension and maximum separation is achieved at ~ pH 5.

441012 Determination of Some Water-Soluble and Fat-Soluble Vitamins in Tears and Blood Serum of Infants and Parents By Liquid Chromatography/Mass Spectrometry

Abstract:

A subset of water-soluble vitamins (WSV) and fat-soluble vitamins (FSV) were detected and quantified in human tears and compared with blood serum levels. This is the first demonstration of vitamin detection simultaneously in tears and blood serum levels in infants and parents. Samples were from 15 four-month-old infants and one of their parents. The vitamin levels were compared between tears and serum in a single subject as well as between infants and parents for the same sample type. Further, concentration results were compared against nutritional information. WSV and FSV were detected with a C_{18} analytical column through two HPLC-ESI-MS/MS 16 min and 18 min methods, respectively. H_2O and acetonitrile (ACN) containing 0.1% formic acid were WSV mobile phases while 1:9 H_2O/ACN and methanol containing 5 mM ammonium formate were FSV mobile phases. Vitamins B_1 (thiamine), B_2 (riboflavin), B_3 (nicotinamide) and E (α-tocopherol) were routinely detected in tears and serum while A (retinol) was detected in serum. WSVs B_5 and B_9 were also detected in tears and blood serum samples however the recovery and precisions found not to be sufficiently reliable. LODs determined in real tears were 0.075-1.7 ng for WSV and 0.24 ng for vitamin E. In contrast, LODs determined in serum were 0.057-0.66 ng for WSV and 1.4 and 0.43 ng for FSVs A and E, respectively, above the clinically reported normal range concentrations. The assays precision and accuracy were validated with %RSD of 2.2-10% and recoveries of 84.0-109.7% for all vitamins. Results demonstrated higher mean values for WSV levels in tears in compared to serum and also in infants in compared to parents. FSV E levels were lower in tears than serum with not significant difference between infants and parents levels while vitamin A was higher in parents than infants. Strong positive correlations were found between tear vs. serum vitamin E for both infants and parents while a slight positive correlation was detected for parent tears vs. serum B_1. Correlations between infant vs. parent were very strong for tears vitamins E and B_2 and serum vitamins E and B_2. A slight correlation was found for vitamin B1 in infant vs. parent tears and serum. Vitamin levels were also cross-correlated with results from self-reported nutritional questionnaires. Results suggest, with additional optimizations to results presented herein, that tears may be a viable biofluid to monitor for nutritional health assessments to speed deficiency diagnoses in infants and parents. Advantages to tears include less invasive sample collection and simplified sample preparation for analysis.

NON-NEWTONIAN FLUID EFFECT ON EFFECTIVE VELOCITIES IN ELECTRICAL FIELD FLOW FRACTIONATION

Stephen Dueck[1] and Mario Oyanader[1]

1. California Baptist University, Riverside, California

Abstract

In of electrical field flow fractionation (EFFF) several parameters have investigated to optimize solute separation [Pascal, and :e, 2010]. Although, devices with a cylindrical geometry the t setting to analyze the effect of couette flows in such as separation of pharmaceuticals and metabolites, the -Newtonian ıre of the carrier fluid has not been fully investigated yet. The main focus of this contribution is to extend the and understanding of the carrier fluid effect on the optimal time of separation. Furthermore, the study concentrates on the of the :al solution using area averaging methodology Newtonian Non-Newtonian cases. Several cases of pseudo- and ıre examined and its influence on the effective velocity . Especial attention is given to fact that traditional simplification of the indrical to a rectangular system does not facilitate the capturing of the Non-Newtonian behavior of fluid.

Process Sketch and Diagrams

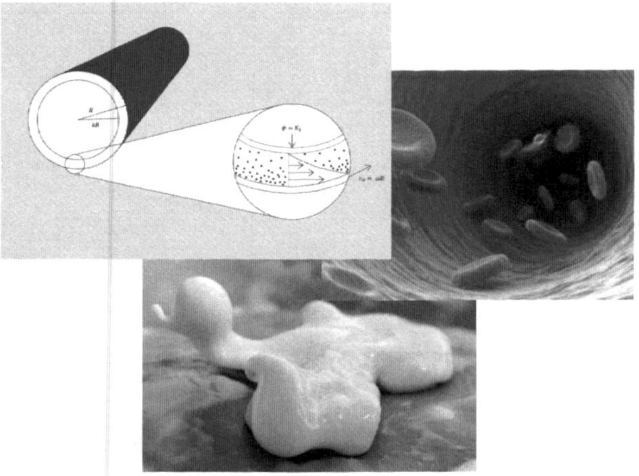

Equations

Carrier Velocity Governing Equation

$$\frac{\partial}{\partial r}(r^2 T_{r\theta}) = 0$$

Shear Stress Power Law Model

$$T_{r\theta} = -m\left(r\frac{\partial}{\partial r}\left(\frac{v_\theta}{r}\right)\right)^n$$

Boundary Conditions

$$v_\theta = 0 \ @ r = kR$$
$$v_\theta = \omega R \ @ r = R$$

Electrical Field Governing Equation

$$\frac{1}{r}\frac{\partial}{\partial r}\left(r\frac{\partial \phi}{\partial r}\right) = 0$$

Boundary Conditions

$$\phi = 0 \ @ r = kR$$
$$\phi = K_1 \ @ r = R$$

Species Continuity Governing Equation

$$\frac{\partial C}{\partial t} + \frac{v_\theta}{r}\frac{\partial}{\partial \theta} = D\left[\frac{1}{r}\frac{\partial}{\partial r}\left(r\frac{\partial C}{\partial r}\right) + \frac{1}{r^2}\frac{\partial^2 C}{\partial \theta^2}\right] + \mu^*\frac{1}{r}\frac{\partial}{\partial r}\left(rC\frac{\partial \phi}{\partial r}\right)$$

Boundary Conditions

$$D\frac{\partial C}{\partial r} + \mu^*(C * \frac{\partial \phi}{\partial r}) = 0 \ @ r = kR$$
$$D\frac{\partial C}{\partial r} + \mu^*(C * \frac{\partial \phi}{\partial r}) = 0 \ @ r = R$$

Effective Velocity

$$\hat{v}_{eff} = 1 + A_0 + A_1 * f$$

$$A_0 = \frac{2k^2 - 2k^{\frac{2}{n}}}{\frac{2}{n} - 2 - \frac{2}{n}k^2 + 2k^{\frac{2}{n}}} \qquad A_1 = \frac{\left(\frac{2}{n} - 2\right)(k^2 - 1)k^{\frac{2}{n}}}{\frac{2}{n} - 2 - \frac{2}{n}k^2 + 2k^{\frac{2}{n}}}$$

$$f = \frac{\left(k^{2+\frac{\Omega}{\ln(k)}} - k^{\frac{2}{n}}\right)\left(2 + \frac{\Omega}{\ln(k)}\right)}{k^{\frac{2}{n}}\left(k^{2+\frac{\Omega}{\ln(k)}} - 1\right)\left(2 + \frac{\Omega}{\ln(k)} - \frac{2}{n}\right)}$$

$$\Omega \equiv \frac{\mu * K_1}{D}$$

Illustrative Cases

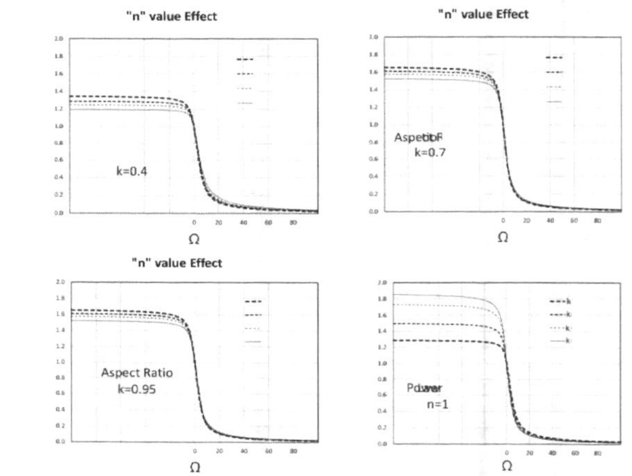

Conclusions

- Effective Velocity values increase as the Aspect Ratio parameter (k) increases.
- Values of the Power Law model parameter "n" prc velocity values as it decreases.
- Inversion of the electrical Field Direction could bring Effective Velocity from maximum values to a minimum of zero.

References

Pascal, J., Oyanader, M. A. and Arce, P., Efffect of Wall Velocities on the Determination of Optimal Separation Times in Elec Fractionation, The Canadian Journal of Chemical Engi . 384-391, 2010.

Oyanader, M. A. and Arce, P., Role of Geometrical Dimensions In Electrophoresis Applications With Orthogonal Fields, pp. 2857-2866, 2005.

Bird, R. B., W. Stewart, and E. N. Lightfoot, *Transport* New York, 2nd Edition, 2002.

cbu | Gordon and Jill Bourns College of Engeneering

AIChE The Global Home of Chemical Engineers

443179 Optimizing Neural Stem Cell Sorting with Dielectrophoresis

Abstract:

Human neural stem and progenitor cells (NSPCs) have therapeutic potential to treat neurodegenerative diseases since they provide neuroprotection and differentiate into the three cell types of the central nervous system - astrocytes, neurons, and oligodendrocytes. However, cultures of these cells are heterogeneous, containing progenitor cells with distinct differentiation properties and little is known regarding which types of progenitors are best for neural repair. Dielectrophoresis (DEP) is a technique that uses alternating current (AC) electric fields to separate cells based on the dielectric properties of their membrane and cytoplasm and has recently shown promise for separating progenitors in the neural lineage.

Using DEP, our group has demonstrated that astrocyte progenitor (AP) and neuron progenitor (NP) cells can be isolated from mouse NSPCs at select frequencies and this behavior is linked to the cells' membrane capacitance and fate potential. We also discovered that NSPCs DEP response was not always the same at a single frequency, which can be attributed to variations in cell populations, microdevices, or media conductivity. To remedy these variations, we've implemented a two-step sorting scheme for NSPCs, which can be extended to other cell systems. Step one defines a cell sample specific trapping frequency, and step two separates cells at that specific frequency in a DEP well microdevice.

In this work human AP cells were detected and separated from heterogeneous NSPCs using DEP. A trapping frequency specific to AP cells ranging from 8-20kHz was defined using a 3DEP reader. Then, a multi-well DEP device with Ti-Au interdigitated electrodes was used to sort AP cells at the trapping frequency. Prior to sorting, a mini-trapping curve was developed over the frequency range 5-100kHz in the multi-well DEP device, confirming acceptable accuracy of the predicted trapping frequency. For post-sorting analysis, cell differentiation and 3DEP reader analysis were completed to verify AP separation, determine cell enrichment, and measure membrane capacitance values of sorted cells. Cells trapped at 1MHz and incubated in the DEP buffer solution were tested as controls. AP separation was confirmed with GFAP immunostaining to detect astrocytes differentiated from APs and the cell enrichment was found to be 1.75 fold. Effectively sorting human NSPCs is essential to further study these cells as treatment options for neurodegenerative diseases.

443180 *A Proposed Method to Isolate, Concentrate, and Identify Bacteria By Dielectrophoresis and Raman Spectroscopy*

Abstract:

Typical bacterial identification methods can take several days to complete. This lengthy analysis time is unacceptable for emergency situations and for life threatening illnesses. In order to reduce analysis time, many researchers have used a variety of methods for bacterial identification such as polymerase chain reaction, Raman spectroscopy, fluorescent in situ hybridization, and micro-array testing. Although these methods have successfully decreased bacterial analysis time from days to a matter of hours, they require a pure sample or a way to label bacteria with fluorescent tags, antibiotics, or primers. Pure samples cannot be obtained directly from patients and appropriate fluorescently marked antibodies increase costs and may result in wasted materials due to the broad range of bacteria strains that cause infections and disease. As such, a label-free identification method is appealing to cut costs and increase simplicity and efficiency. One label-free method is dielectrophoresis (DEP), which is the phenomenon by which motion in a particle occurs as it passes through a non-uniform electric field. The motion is a result of the applied frequency of the electric field and the particles' intrinsic properties. This study proposes to use DEP as a means to isolate and concentrate bacteria for subsequent identification through Raman spectroscopy and statistical classification methods. If successful, the method could be used to identify bacteria from a mixed sample in a matter of minutes.

444644 Selective Trapping and Enrichment of Particles Reversing the Elution Order in Insulator Based Dielectrophoresis

Abstract:

Insulator-based dielectrophoresis (iDEP) is a commonly used microfluidic technique for studies of enrichment and separation of samples with biological cells. Additionally, iDEP lab-on-a-chip systems have been demonstrated to be highly useful in areas where rapid results are favorable, such as food/water safety, and clinical diagnostics. In the past, these systems have operated by eluting from the system small particles first. In the case of biological samples where large particles are targeted, a system in which these larger cells can be eluted first is, consequently reducing the amount of time they are exposed to the possible negative effects of iDEP, is needed. The present study focuses on an reversed elution order technique, in which particles from largest to smallest in tertiary mixtures will be enriched, selected, and separated into different outlets, seeking to minimize the time to which the various fragile biological cells are exposed to negative dielectrophoretic effects.

429282 Conductivity Theory Fitting Measured Data for Various Alcohol-Toluene Mixtures Across Entire Concentration Range

Abstract:

We have measured the conductivity of binary alcohol-toluene mixtures (using 8 different alcohols) across the entire concentration range. The conductivity varies from 10^{-11} up to 3.10^{-4} S/m and, for a certain range decays rapidly and non-linearly with decreasing alcohol content. We assume that this range is associated with ion-pair formation due to decrease in dielectric constant. The Onsager-Fuoss conductivity theory is unable fit experimental conductivity across such wide range. Because of this we have suggested a more general conductivity theory that is based on the known expression for ionic strength, which has been derived directly from mass-action law for ion/ion-pair equilibrium (Aurbach [1999]). This theory fits experimental conductivity for all mixtures, for data which spans over 7 orders of magnitude. The adjustable parameters in this general theory are total ion concentration in the pure alcohol, the shortest distance of ion approach in the ion-pair, and hydrodynamic size of the ions.

434537 Electrochemical Impedance Spectroscopy of Doped Nonpolar Liquids II: Adsorption of Charge Carriers

Abstract:

Electrochemical impedance spectroscopy of doped nonpolar liquids II: adsorption of charge carriers

AIChE Annual Meeting 2015, Salt Lake City, UT

Session T3010: Electrokinetics in Non-Polar Media

Benjamin A. Yezer, Aditya S. Khair, Paul J. Sides and Dennis C. Prieve

Center for Complex Fluids Engineering

Department of Chemical Engineering

Carnegie Mellon University

Pittsburgh, PA 15213

Abstract

Surfactants are added to nonpolar media to increase the charge concentration and charge solid surfaces. Electrochemical impedance spectroscopy has been used to determine the permittivity, charge carrier concentration, and charge carrier size of solutions of dodecane doped with different surfactants.[1] The spectra showed the frequency response of various phenomena within the cell, such as the geometric charging, conductance, and double layer charging. The impedance of dodecane doped with OLOA 11000 [poly(isobutylene succinimide)], Span 80 (sorbitan monooleate), and Span 20 (sorbitan monolaurate) was measured as a function of excitation by a 10 mV amplitude sinusoidal voltage applied across a parallel plate cell with a 10 micron spacing. The tested solutions varied in concentration from 1 mM to 100 mM and the frequency range was 10 mHz to 100 kHz. Nyquist plots of all three surfactants showed the high frequency semicircle characteristic of parallel resistance and capacitance. The plots of Span 80 and Span 20 in dodecane also exhibited a second semicircle at low frequencies. The electrical conductivity of each surfactant was proportional to surfactant concentration for concentrations above 10 mM. The low frequency behavior was attributed to charge adsorption and desorption. Fitting the data to models for charge migration, differential capacitance, and adsorption allowed extraction of both charge concentration and two kinetic parameters that characterize the rate of adsorption and desorption. Above 10 mM the ratio of charge carriers per surfactant molecule was 17 ppm for OLOA 11000, 23 ppm for Span 20, and 3 ppm for Span 80. Larger micelles had an exponentially higher ratio of charge carriers per molecule of surfactant. The adsorption rate constants were independent of surfactant concentration while the desorption rate constant was proportional to the surfactant concentration when the concentration was above the critical micelle concentration (CMC). This dependence indicated that uncharged micelles participated in the desorption of charge.

---

[1]. B.A. Yezer, A.S. Khair, P.J. Sides and D.C. Prieve, "Use of electrochemical impedance spectroscopy to determine double-layer capacitance in doped nonpolar liquids," *J. Colloid Interf. Sci. 2015* ***449****, 2-12 (doi:10.1016/j.jcis.2014.08.052).*

431875 Transition from Steric to Electrostatically Stabilized Carbon Black Suspensions in Nonpolar Solvents

Abstract:

Stabilization of suspensions in nonpolar media by charge transfer and generation of long range electric double layer interactions has been proposed. The development of stabilizers and dispersants for particle suspensions in nonpolar solvents often involves the characterization of large composition space given a lack of understanding of the details of the stabilizing mechanism. Using microfluidic and millifluidic approaches to quantify the timescales and lengthscales of sedimentation and stabilization in nonpolar solvents, we have characterized the transition from steric to electrostatic stabilization in a well-studied system; carbon black stabilized with polyalkyl succinic acid. A shift in the concentration of stabilizer needed to achieve long term stability is observed with change in solvent. To connect this shift to mechanism, the impact of solvent on micelle formation, adsorption to the particle interface and sedimentation has been quantified. Results allow us to investigate the proposed stabilization mechanism by varying composition and solvent.

430233 The Acid-Base Mechanism for the Charging of Colloidal Particles in Apolar Media

Abstract:

SourceURL:file://localhost/Users/johnberg/Desktop/AIChE%20Abstract.doc @font-face { ""; }p.MsoNormal, li.MsoNormal, div.MsoNormal { margin: 0in 0in 0.0001pt; font-size: 12pt; ""; }table.MsoNormalTable { font-size: 10pt; ""; }div.Section1 { page: Section1; }

The charging of colloidal particles in apolar media (dielectric constant ≈ 2) is a subject of much current interest, underlying such applications as electrophoretic displays (e.g., the Amazon Kindle® reader) and new printing devices (e.g., the HP Indigo® Digital Press). It is often of interest to charge a particular particle type in a given medium with a given polarity and magnitude, and for this purpose it is desirable to have sufficient mechanistic understanding of the charging process to provide guidelines for the creation of the desired result. Understanding the creation and stabilization of electric charge in apolar media, and in particular the charging of colloidal particles with them has been challenging owing to the complexity of the systems involved and the large number of factors that appear to be important. It is generally accepted that charges in measurable amount can exist in apolar media only when housed in reverse micelles, or similar structures, composed of appropriate surfactant monomers, and that small numbers of charges may arise in such systems by micelle-micelle disproportionation. It is likely that the charge-bearing species in the cores of the micelles are hydroxyl and hydronium ions arising from traces of water housed there. With respect to the charging of particle surfaces in apolar media, however, a number of different, and sometimes conflicting, theories have been advanced. It appears now that most observations, obtained both in the author's laboratory and elsewhere, are explainable in terms of an acid-base mechanism, and the present paper reviews this evidence. Adducts formed between chemical functional groups on the particle surface and monomers of reverse micelle-forming surfactants dissociate, leaving charged groups on the surface, while the counter-charges formed are sequestered in the reverse micelles. For a series of mineral oxides in a given medium with a given surfactant, surface charging (as quantified by the maximum electrophoretic mobility or zeta potential obtained as surfactant concentration is varied) was found to scale linearly with the aqueous *PZC* (or *IEP*) values of the oxides. Different surfactants, with the same oxide series, yielded similar behavior, but with different *PZC* crossover points between negative and positive particle charging, and different slopes of charge vs. *PZC*. Thus the oxide series could be used as a yardstick to characterize the acid-base properties of the surfactants. This has led directly to the study of other materials, including surface-modified oxides, carbon blacks, pigments (charge transfer complexes), and polymer latices. This review focuses on the acid-base mechanism of particle charging in the context of the many other factors that are important to the phenomenon, including the presence of water, of other components (e.g., synergists, contaminants, etc.) and of electric field effects. The goal is the construction of a road map describing the anticipated particle charging behavior in a wide variety of systems, assisting in the choice or development of materials for specific applications.

436375 Mechanisms of Electric Charging By Surfactants in Nonpolar Dispersions

Abstract:

We present electric charging phenomena caused by surfactants in nonpolar media, and apply small variations in the surfactant composition to identify some of the physicochemical parameters crucial for the charging process. Although we observe striking differences to the more familiar charging in aqueous media, our experiments nonetheless support the notion that even in nonpolar systems water plays an important and multi-facetted role for charge formation. For the charging of polymer particles in nonpolar solutions of 'humid' surfactant micelles, we discuss a hypothetical two-step mechanism involving the formation and surface adsorption of micellar ions with substantial size asymmetry.

427847 Contact Charge Electrophoresis: Experiment and Theory

Abstract:

Contact charge electrophoresis (CCEP) uses steady electric fields to drive the continuous, oscillatory motion of conductive particles and droplets between two or more electrodes. These rapid oscillations can be rectified to direct the motion of objects within microfluidic environments using low-power, dc voltage. Here, we compare high precision experimental measurements of CCEP within a microfluidic system to equally detailed theoretical predictions on the motion of a conductive particle between parallel electrodes. We use a simple, capillary microfluidic platform that combines high-speed imaging with precision electrical measurements to enable the synchronized acquisition of both the particle location and the electric current due to particle motion. The experimental results are compared to those of a theoretical model, which relies on a Stokesian dynamics approach to accurately describe both the electrostatic and hydrodynamic problems governing particle motion. We find remarkable agreement between theory and experiment, suggesting that particle motion can be accurately captured by a combination of classical electrostatics and low-Reynolds number hydrodynamics. Building on this agreement, we offer new insight into the charge transfer process that occurs when the particle nears contact with an electrode surface. In particular, we find that the particle does not make mechanical contact with the electrode but rather that charge transfer occurs at finite surface separations of >0.1 μm by means of an electric discharge through a thin lubricating film. We discuss the implications of these findings on the charging of the particle and its subsequent dynamics.

UNIVERSITY OF NOTRE DAME

Fluidic Resistive Random-Access Memories (FRRAM) for Multiplex Nanomembrane Sensor Arrays

Gongchen Sun, Zdenek Slouka, Satyajyoti Senapati, Hsueh-Chia Chang

University of Notre Dame
Dept. of Chemical and Biomolecular Engineering

1

Tue. Nov. 10, 2015
2:15pm – 2:30pm
Salt Palace CC
Ballroom E

Ion current nanomembrane sensor for multi-target biosensing

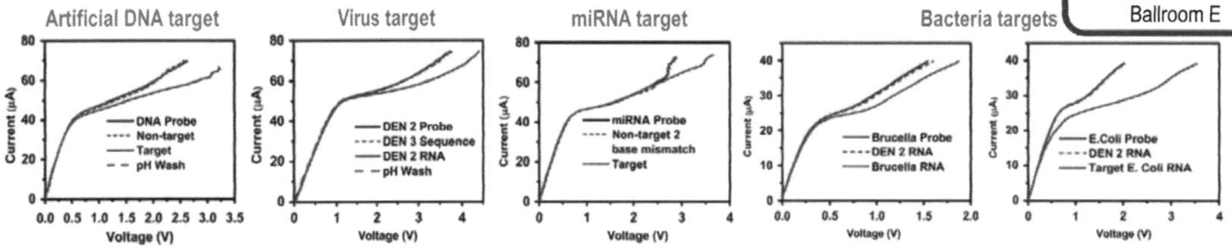

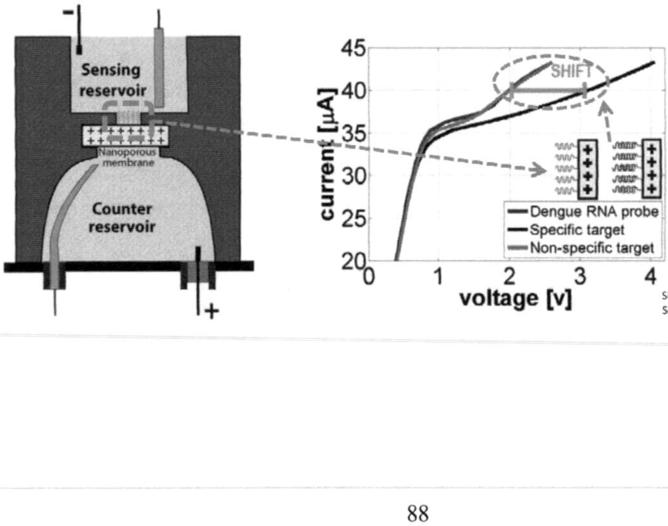

Voltage signature shifts as target molecules hybridize onto membrane surface

Slouka Z. et al, Annu Rev Anal Chem, 7, 317-335, 2014
Senapati S. et al, Biosens Bioelectron, 60, 92-100, 2014

2

Ion current nanomembrane sensor for multi-target biosensing

Tue. Nov. 10, 2015
2:15pm – 2:30pm
Salt Palace CC
Ballroom E

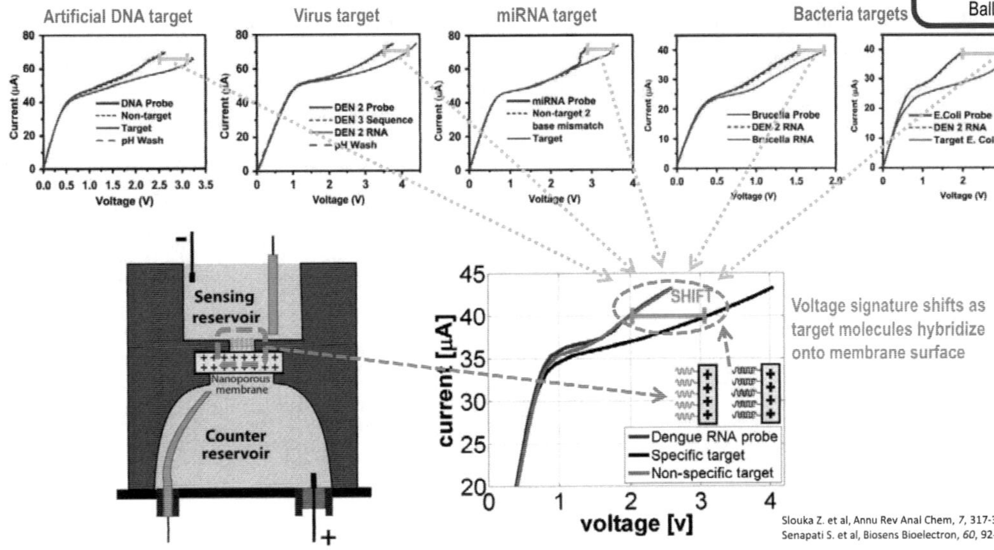

Slouka Z. et al, Annu Rev Anal Chem, *7*, 317-335, 2014
Senapati S. et al, Biosens Bioelectron, *60*, 92-100, 2014

3

Integration of multiplex ion current nanomembrane sensor platform

The Need

→ Register and memorize the voltage shift from membrane sensors (Store the sensing result for further analysis)
→ Facilitate a parallel sensing paradigm for rapid detection of a large library of targets

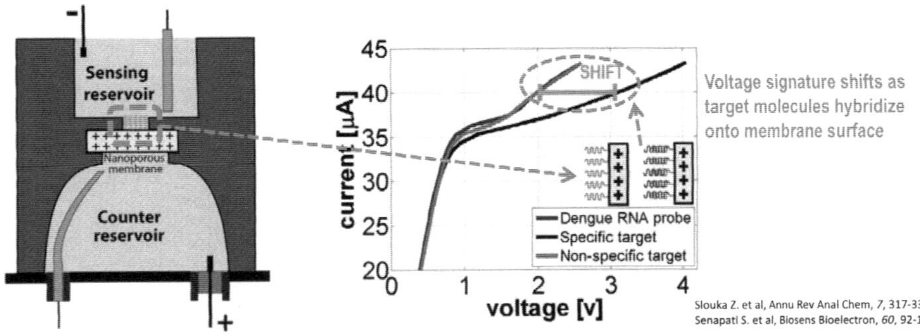

Slouka Z. et al, Annu Rev Anal Chem, *7*, 317-335, 2014
Senapati S. et al, Biosens Bioelectron, *60*, 92-100, 2014

4

Memristor --- transform voltage signatures into a non-volatile form → resistance

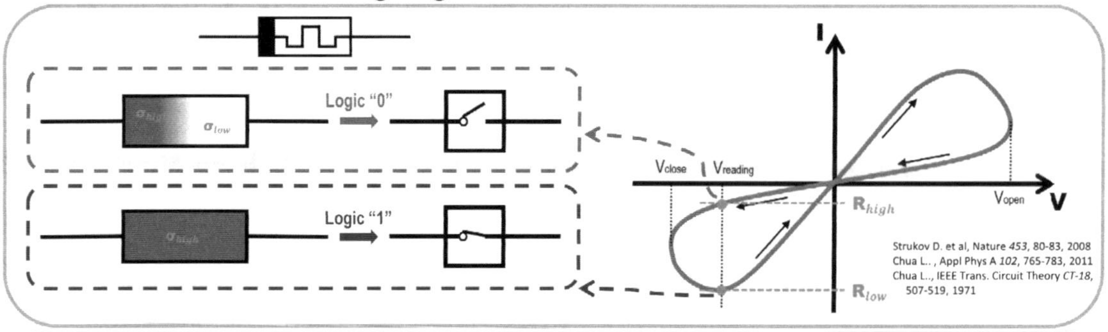

Memristor (memory+resistor):

→ Defined by a state-dependent Ohm's Law:

$$v = R(w)i$$

$$\frac{dw}{dt} = i$$

"w" is a state variable of a particular device, changed when current flow through.

Fingerprint (experimental criteria) of a memristor: Pinched hysteresis loop when driven by bipolar periodic voltage v(t) or current i(t)

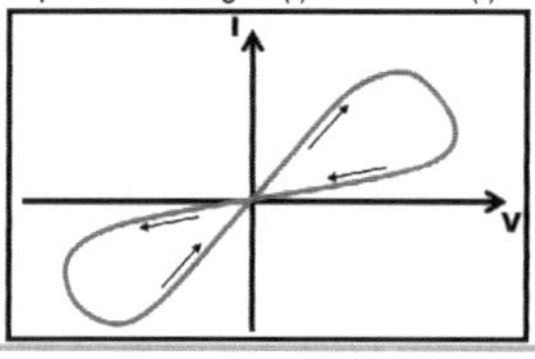

→ Non-volatile memory arrays

→ Elementary block for logic devices

5

Memristor --- transform voltage signatures into a non-volatile form → resistance

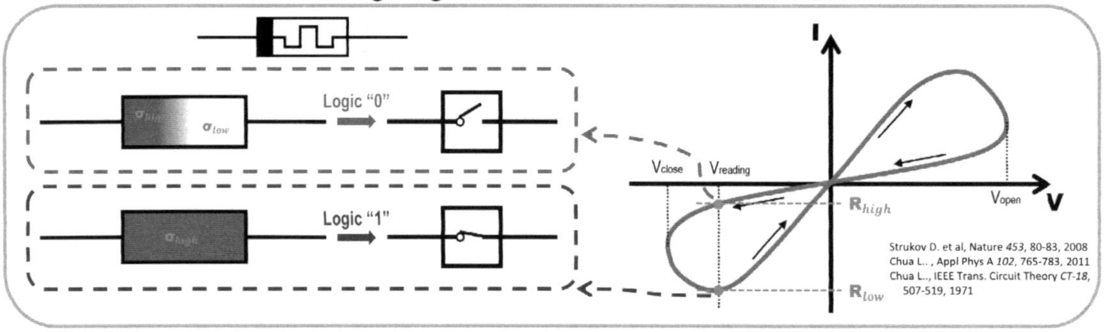

A memristor for fluidic-based ionic system is needed

→ **Transform voltage signatures carried by ion current in electrolytes into resistance**

6

Strategy of designing a fluidic-based ion memristor

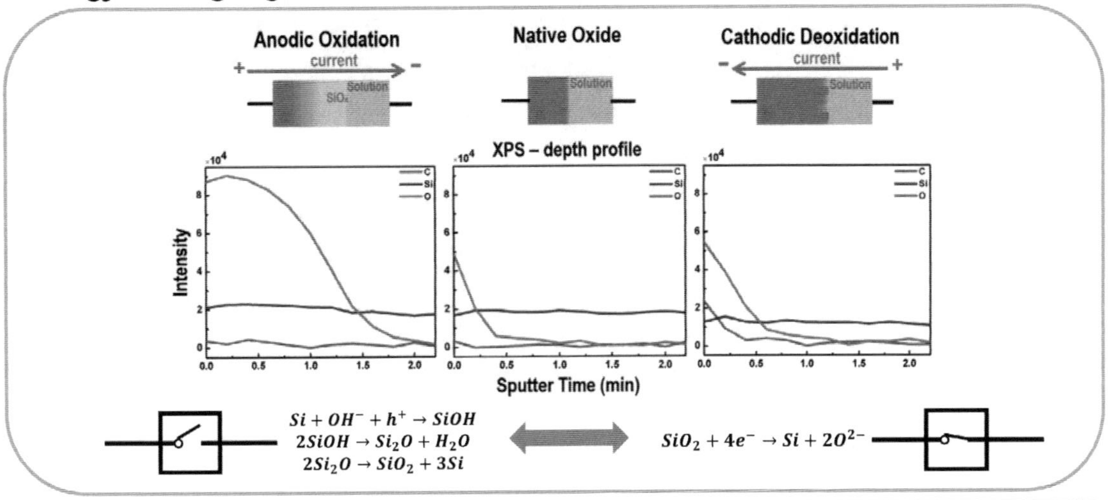

State variable: the formation of interfacial oxide layer on the silicon microelectrode

7

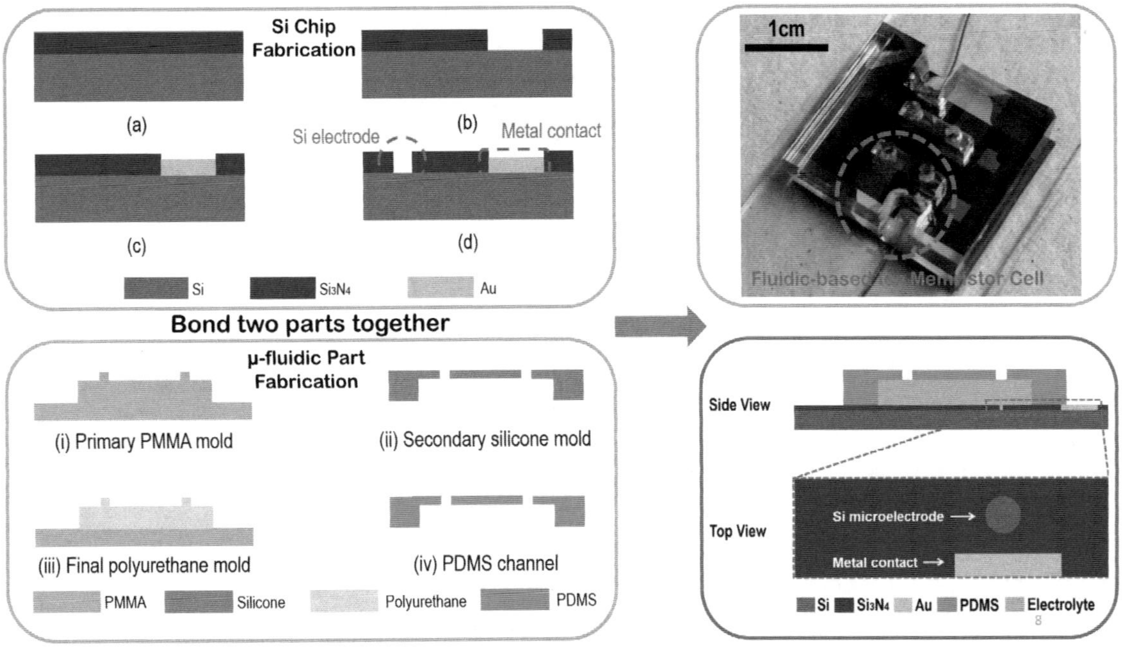

8

Pinched hysteresis fingerprint of the fluidic-based ion memristor

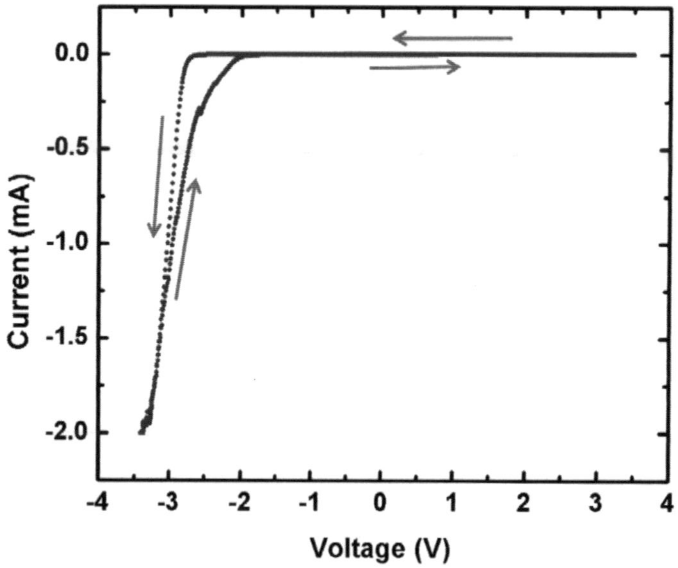

9

Pinched hysteresis fingerprint of the fluidic-based ion memristor

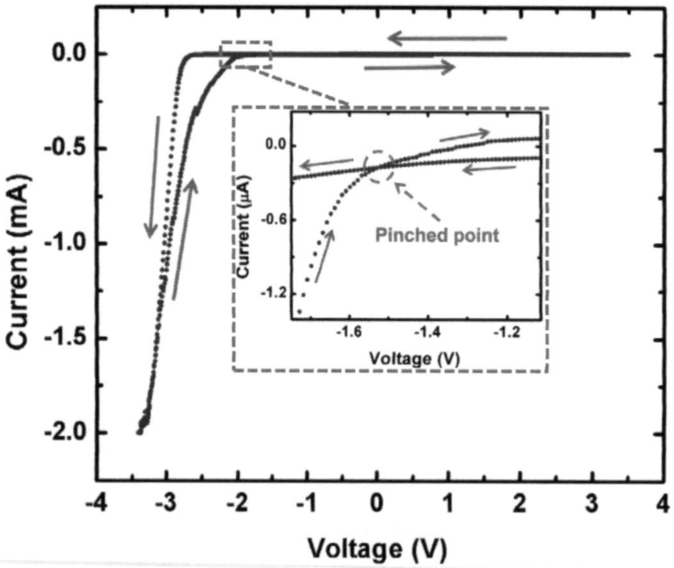

10

Pinched hysteresis fingerprint of the fluidic-based ion memristor

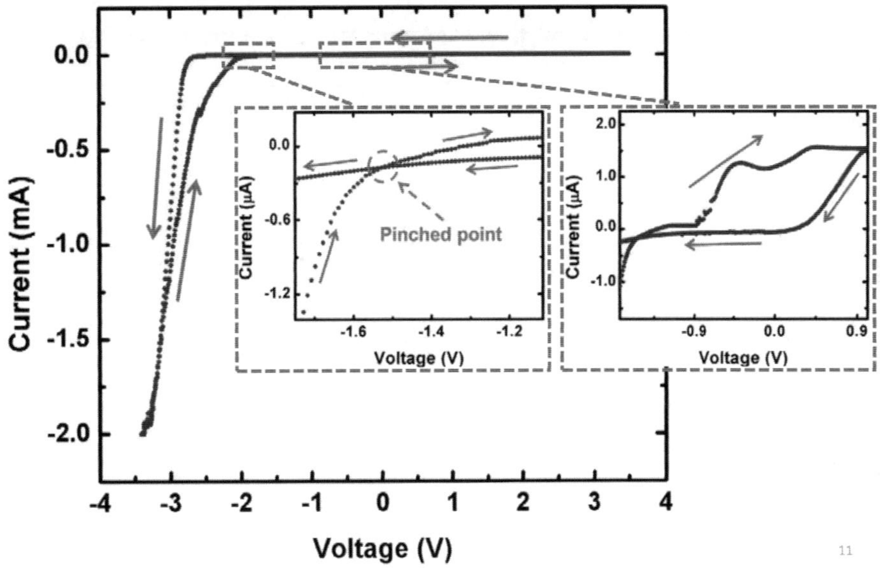

11

Pinched hysteresis fingerprint of the fluidic-based ion memristor

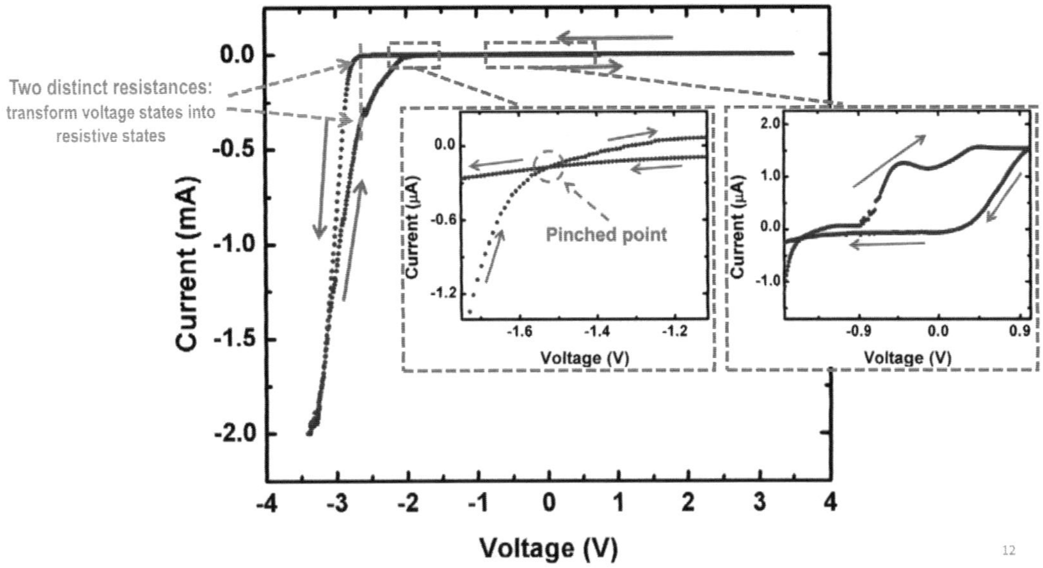

12

Strategy of integrating nanomembrane sensors with the fluidic-based ion memristor

→ Digitizing voltage signatures from sensors as the input of the fluidic-based ion memristor

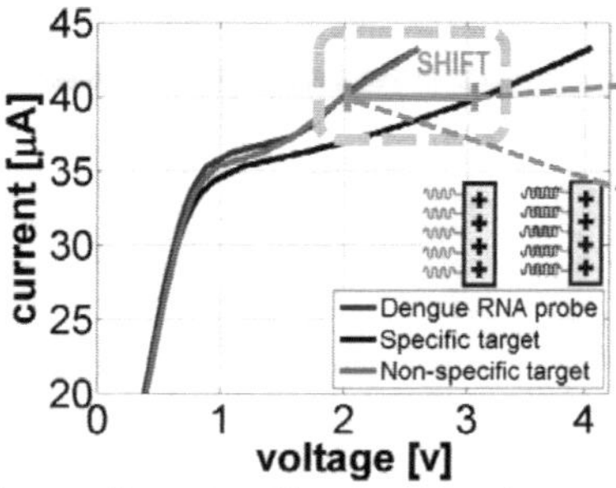

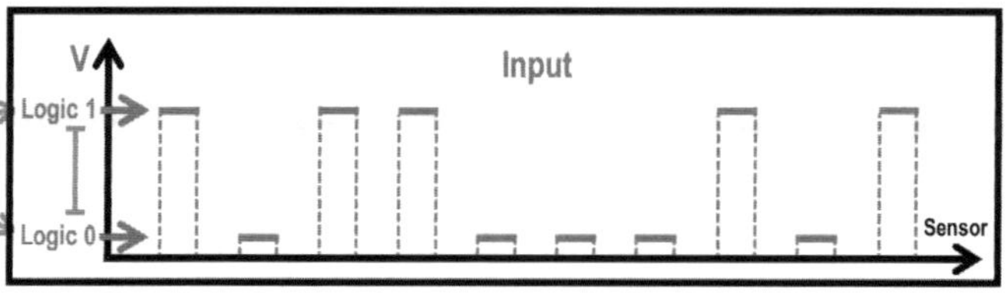

Voltage difference between logic "1" and logic "0" ⟷ Nanomembrane sensor voltage shift due to target hybridization

V_{input}: voltage signal from nanomembrane sensors

Specific target not present → V_{input} ~ Logic "0"
Specific target present → V_{input} ~ Logic "1"

Requirement:
→ Minimum voltage difference between binary logic "1" and logic "0" the fluidic-based ion memristor can register < 1V

13

Fluidic-based resistive random-access memory (FRRAM) cell

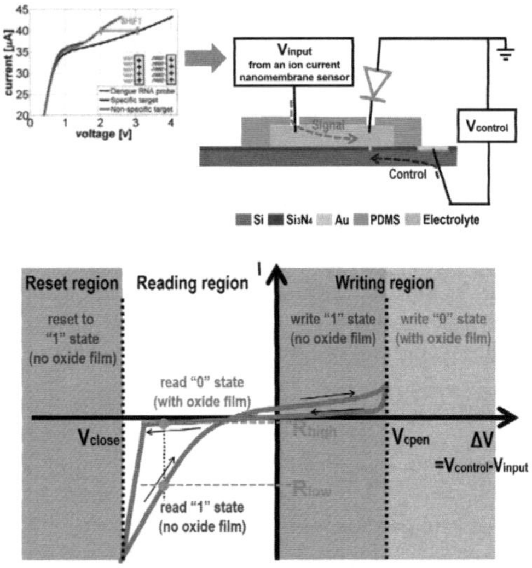

Writing/reading protocol

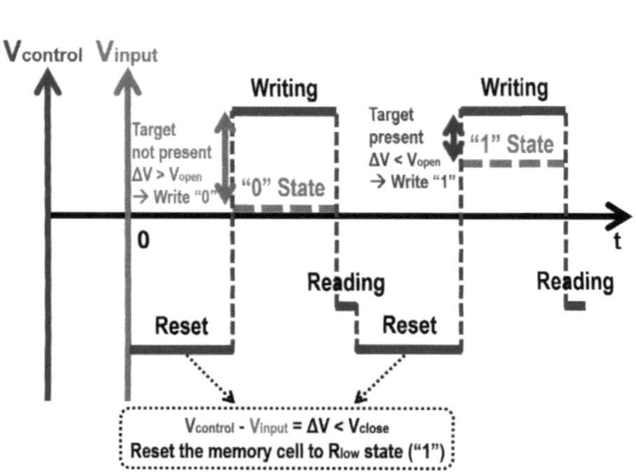

14

Design of the writing protocol – the criteria of choosing $V_{writing}$

On/Off ratio of the fluidic-based ion memristor:

$\frac{R_{high}}{R_{low}}$ @ reading

High On/Off ratio
→ $\Delta V > V_{open}$

Low On/Off ratio
→ $\Delta V < V_{open}$

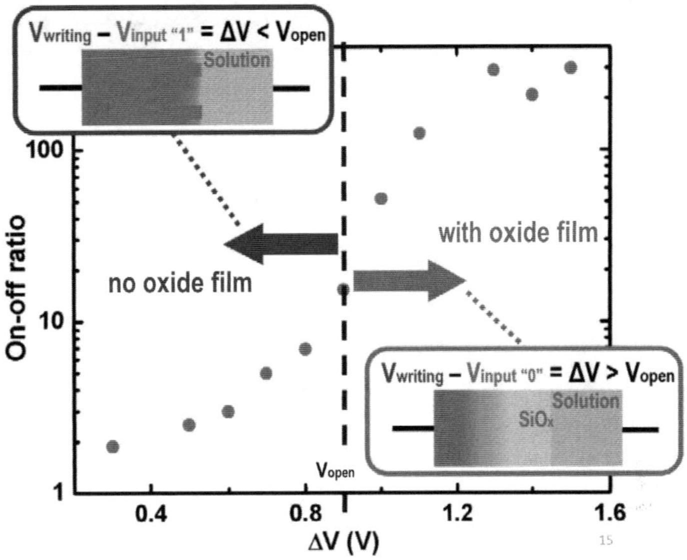

Writing/reading test of a FRRAM cell

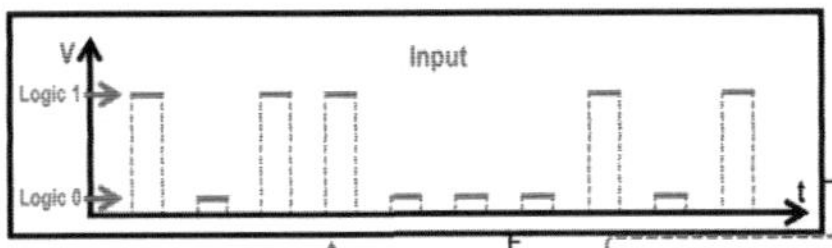

Control:
→ $V_{writing}$ ~ 1.3V

Logic:
→ Logic "1" ~ 1.3V
→ Logic "0" ~ 0V

Latch value:
→ R_{low} ~ "1"
→ R_{high} ~ "0"

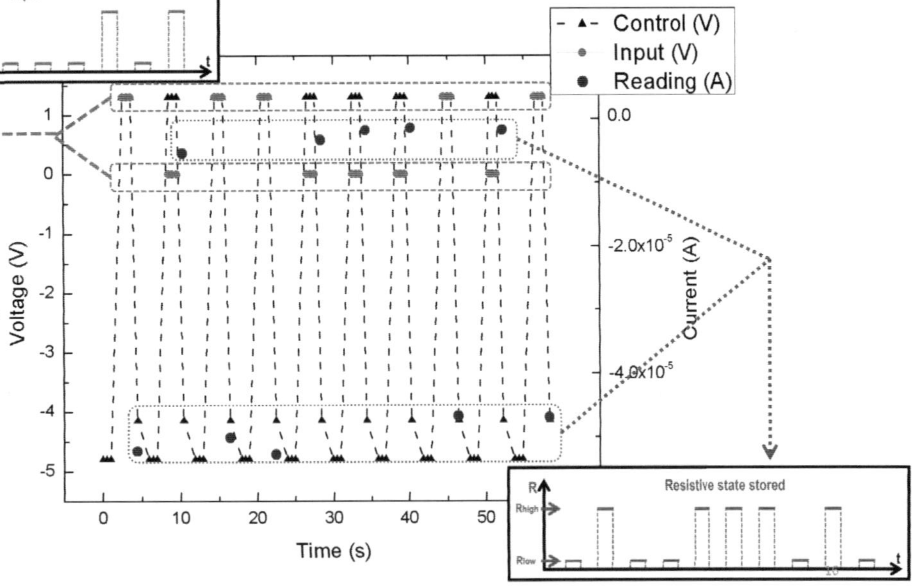

Writing/reading test of a FRRAM cell

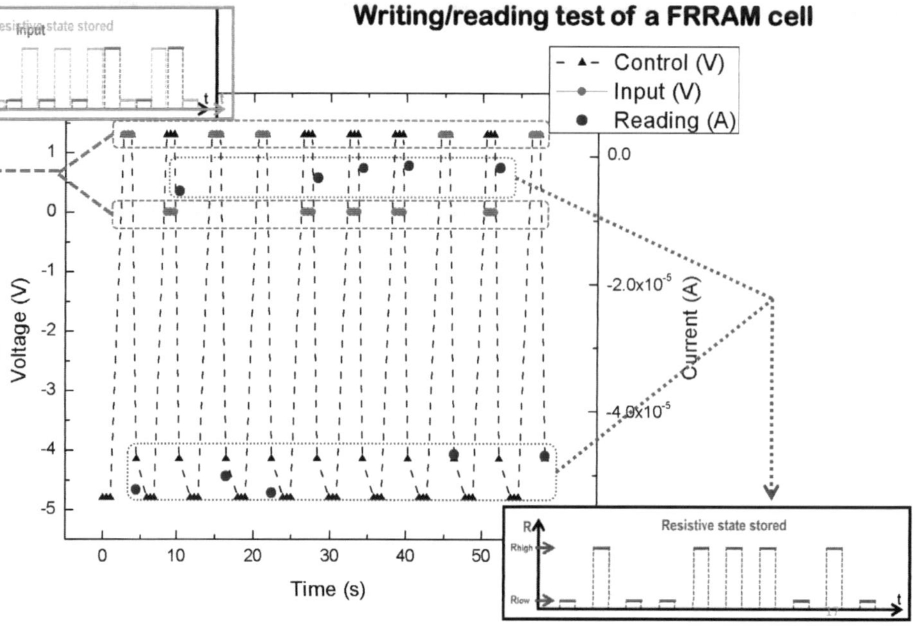

Control:
→ $V_{writing}$ ~ 1.3V

Logic:
→ Logic "1" ~ 1.3V
→ Logic "0" ~ 0V

Latch value:
→ R_{low} ~ "1"
→ R_{high} ~ "0"

Memory cell performance vs Input voltage difference between binary logic values

→ $V_{input\ "0"}$: 0V
→ $V_{input\ "1"}$: 0.3V ~ 1.3V

Results:

→ High On/Off ratio (> 5) achieved for > 600mV difference between binary logic inputs

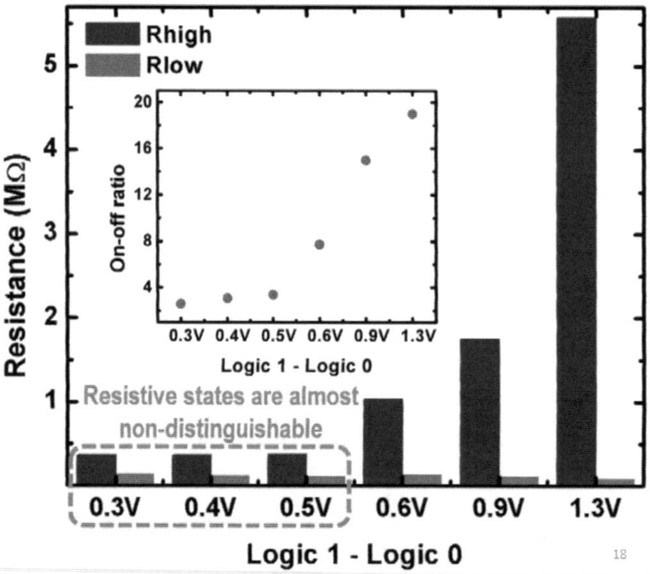

Parallel writing paradigm of FRRAM integrated multiplex nanomembrane sensor array

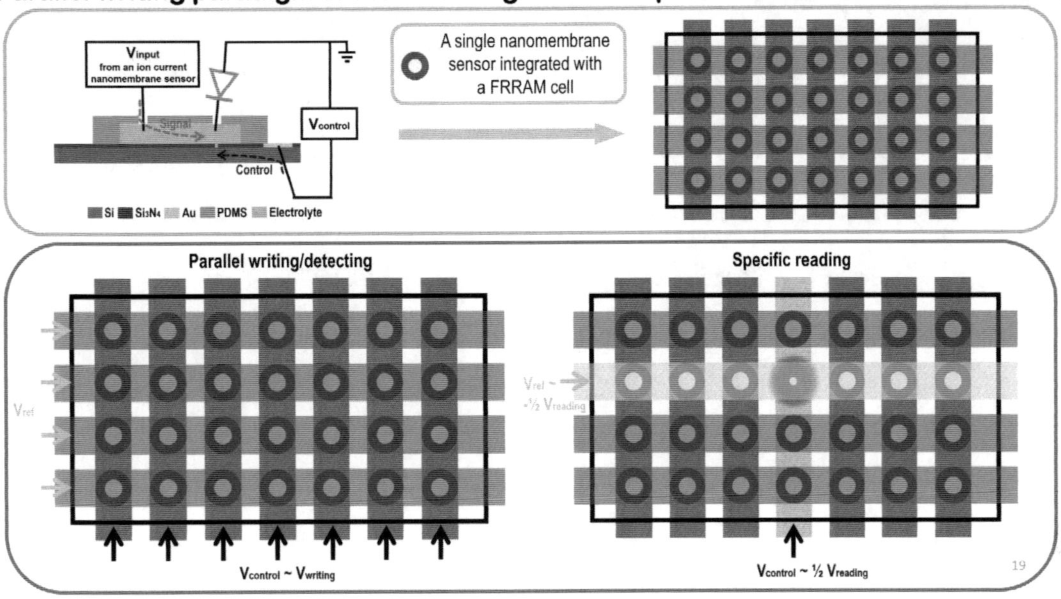

19

Fluidic resistive random-access memories → integrating multiplex nanomembrane sensor arrays

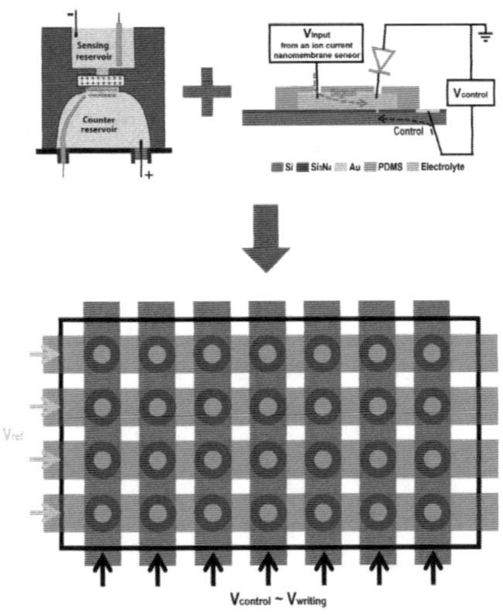

- A fluidic-based ion memristor for transforming voltage signatures from the nanomembrane sensor to non-volatile resistive states
- A fluidic resistive random-access memory (FRRAM) with optimal operation protocol to integrate the nanomembrane sensor and the fluidic-based ion memristor
- A feasible parallel detection paradigm for multiplex nanomembrane sensor arrays

small
Explore this journal >
Communication
Fluidic-Based Ion Memristors and Ionic Latches
Gongchen Sun, Zdenek Slouka, Hsueh-Chia Chang
First published: 6 August 2015 Full publication history
DOI: 10.1002/smll.201501229 View/save citation
View issue TOC
Volume 11, Issue 39
October 21, 2015
Pages 5206–5213

20

Fluidic Resistive Random-Access Memories (FRRAM) for Multiplex Nanomembrane Sensor Arrays

Gongchen Sun, Zdenek Slouka, Satyajyoti Senapati, Hsueh-Chia Chang

University of Notre Dame
Dept. of Chemical and Biomolecular Engineering

21

Transforming voltage states into resistive states

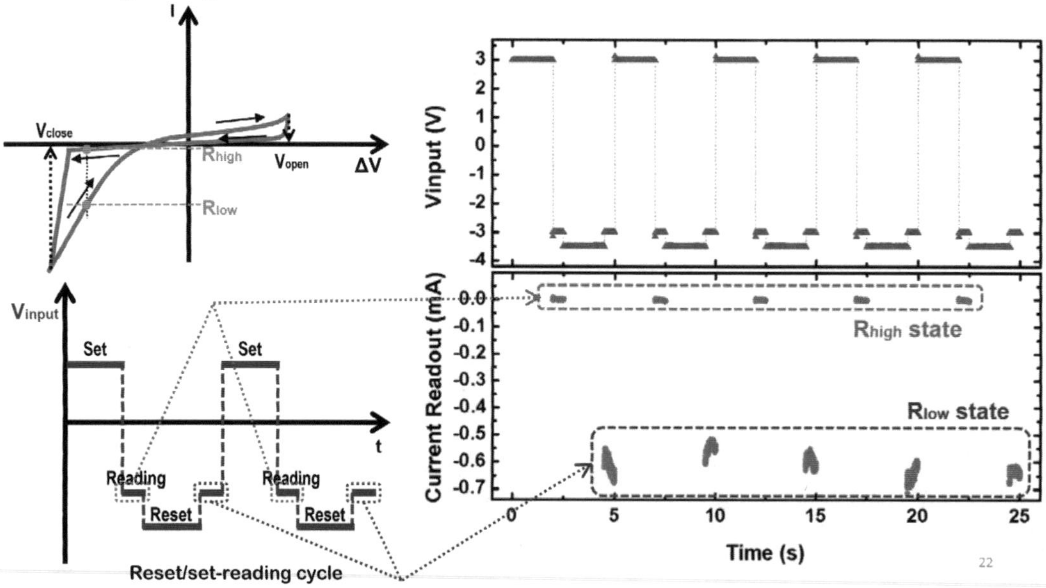

22

436827 Enabling Ultrafast Immunoassays By Dielectrophoretic Biomarker Enrichment on a Nanoslit Device Platform

Abstract:

Heterogeneous immunoassays usually require long incubation times to promote specific target binding and several wash steps to eliminate non-specific binding. Hence, signal saturation is rarely achieved at detection limit analyte levels, leading to significant errors in analyte quantification due to the extreme sensitivity of the signals to incubation time and methodology. Additionally, conventional immunoassay methodologies are not well suited for working with the sub-nanoliter level sample volumes often encountered within point-of-care diagnostic and therapeutic applications, such as for the serial monitoring of cells from liquid biopsies. The poor binding kinetics of immunoassays at detection limit levels can be alleviated through creating a highly concentrated analyte plug that is localized in the vicinity of immobilized capture probes. A commonly investigated methodology for achieving highly concentrated analyte plugs from dilute samples is through electrokinetic enrichment due to ion depletion at the micro/nanochannel interface for coupling with pre-immobilized antibodies. An alternate methodology to enhance biomarker binding kinetics is by immobilizing antibodies inside nano-slit channels, since this device geometry significantly reduces the extent of the diffusion layer. However, no prior work has applied electrokinetic enrichment of the biomarkers inside nano-slit device geometries with immobilized capture probes, to explore the enhancement in analyte binding kinetics obtained by combining these two approaches. Herein, we present a device to achieve this outcome by coupling frequency-selective dielectrophoretic enrichment of the target biomarker in physiological media to capture probes immobilized in a nano-slit channel to enable ultrafast immunoassays that exhibit near-instantaneous (< 2 minutes) signal saturation with dilute biomarker samples (picomolar levels) within ultra-low sample volumes (picoliters). By coupling dielectrophoretic biomarker enrichment with electrochemical detection methods, we are able to eliminate the need for wash steps to remove excess secondary antibody or reporter molecules, since electron transfer rates steeply drop-off for electroactive molecules localized beyond a few nanometers from the electrode surface. Based on the reduced signal variations at early time points with detection limit levels of Prostate Specific Antigen, and the relative insensitivity of the signals to incubation time at later time points, we envision a method for carrying out more accurate immunoassays using analyte enrichment in nano-slits.

References:

B. Sanghavi, W. Varhue, J. Chavez, C.F. Chou, N. S. Swami; "Electrokinetic preconcentration and detection of neuropeptides at patterned graphene-modified electrodes in a nano-channel device", *Anal. Chem.* (2014), 86 (9), pp 4120–4125.

428055 Diffusion-Based Microfluidic Bisulfite Conversion for DNA Methylation Detection

Abstract:

Extensive studies have demonstrated that DNA methylation has played a major role in many biological procedures, such as cellular proliferation and differentiation. Bisulfite genomic sequencing is regarded as the gold-standard technology for analyzing DNA methylation. This technology is based on the different consequences after cytosine and 5-methycytosine (5-mC) treated with sodium bisulfate. The cytosine in DNA will be converted to uracil residues and recognized as thymine in subsequent PCR amplification and sequencing. On the other hand, the 5-mC will remain the same so that it can be distinguished from cytosine.

One of the major concerns of bisulfite conversion is the degradation of DNA. Under known conditions, at least 80% (up to 98%) of the DNA is degraded due to the non-specific degradation under acidic conditions and more significant losses are found to occur during the subsequent removal of bisulfite and desulphonation steps. To obtain decent sequencing coverage in the whole genome, hundreds of nanograms (up to 2 μg) input DNA is needed. This results in huge problem when dealing with rare samples, such as blood sample and embryo. In addition, the bisulfite conversion usually takes 4~16 hours in traditional protocols and commercial available kits. The long incubation time results in high conversion rate, but also significantly affects the integrity of DNA. To reduce the amount of DNA needed, we developed a diffusion-based microfluidic chip to minimize sample loss by decreasing reaction time without compromising conversion rate. In addition, the diffusion strategy eliminates the need for column based DNA purification which improves the sensitivity of the assay. This strategy will be potentially used for reduced representative bisulfite sequencing (RRBS) and whole genome bisulfite sequencing (WGBS) with limited DNA samples.

In this work, we demonstrate a simple scheme for conducting multi-step DNA bisulfite conversion, including denaturation, sulphonation, desalting, desulphonation and elution. To the best of our knowledge, this is the first demonstration of DNA bisulfite conversion on a microfluidic platform. Our microfluidic device has a simple structure that includes a reaction chamber connected with two loading chambers on both sides. The connections between the reaction chamber and the two loading chambers could be cut off by closing two-layer valves. This simple microfluidic platform allows the replacement of small molecules inside the reaction chamber by diffusion from/into the loading chambers under the concentration gradient, while preserve the large DNAs inside throughout such process. This device eliminates the need of column-based purification which results in higher DNA recovery and better DNA integrity. The diffusion scheme also increases the concentration of bisulfite mix which shorten the reaction time from 4~16 hours to 2 hours. The single chamber ("one-pot") design drastically minimizes the complexity of the microfluidic device and improves the performance of DNA bisulfite conversion.

428599 Salt Gradient Regulation of Ion and Nanoparticle Transport in Nanopores with Bipolar Charges

Abstract:

Fundamental understanding of ion and nanoparticle (NP) transport is crucial for designing nanofluidic/nanopore devices and characterizing individual unlabeled (bio)nanoparticles. To enhance the resolution in the nanopore-based sensing device, two critical challenges need to be resolved, including enhance capture rate of NP and reduce simultaneously its translocation velocity through a nanopore. To achieve these, two types of bipolar nanopore, namely, only the inner surface (case I) and both the inner and outer surfaces (case II) of the nanopore with bipolar charges, are investigated. Results show that the ion current rectification (ICR) can be significantly regulated by a salt concentration gradient and even a reversed trend is observed when the salt gradient is large enough. Furthermore, we demonstrate that NP can be controlled effiecently in the bipolar nanopores. The simulation results gathered suggest that applying a salt gradient to a case I nanopore has the potential in single molecules sensing because it is capable of raising both the capture efficiency and the sensing resolution.

431134 Performing Gene Mapping on DNA Unraveled on Patterned Lipid Bilayers

Abstract:

We have developed a new platform for stretching DNA and performed rapid gene mapping on the platform using various labeling methods. In this study, DNA were first adsorbed on cationic-lipid membranes that are supported on grooved glass surface. The adsorbed DNA then spontaneously gathered and unraveled along the grooves. For both T4GT7(166kbp) and lambda DNA(48.5kbp), their relative extension can reach as high as 85%, comparable to the value obtained as DNA were confined in sub-50nm nanochannels. On this platform, we have performed gene mapping on both lambda and T4GT7 DNA using several labeling approaches including restriction digestion, fluorescent peptide nucleic acids, and nick translation. The locations of labeling agents (or the corresponding sequences) on DNA were determined using fluorescence microscopy. All three labeling methods were found compatible with the lipid platform and their results are at similar degree of accuracy. Since our method is faster and cheaper than the traditional ones, it has great potential for future gene mapping applications.

433657 An Integrated Diagnostic Platform for Biomolecule Detection

Abstract:

Current techniques for detection of biomolecules including DNA, RNA and lipids are expensive and require trained personnel/extensive sample prep/bulky instrumentation, making it impractical for field applications. The development of low-cost, turn-key and portable detection platforms for identification of biomolecules has attracted great attention to a wide variety of applications in food, water and medical safety. We report an integrated, low-cost, sample-to-answer, and selective biosensing platform for the detection of biomarkers. The integrated microfluidic biochip consists of a sample pretreatment unit to isolate biomolecules from cells/bacteria/virus in raw samples and a nanomembrane-based molecular sensing/pre-concentration unit to perform the biosensing. The sample pretreatment unit is fabricated based on gel electrophoresis where an agar gel is used to separate negatively-charged biomolecules from a target sample by applying a DC field. Downstream from the pretreatment unit, an ion-selective nanomembrane-based preconcentration unit is used to concentrate the isolated molecules at a specific position where the biosensing unit is placed. The concentration of biomolecules (up to 200X) is due to the depletion phenomenon developed at the cation-exchanged membrane interface and application of pressure driven flow in the opposite direction of the depletion front. The anion-exchange nanomembrane-based biosensing is based on a charge inversion phenomenon that occurs when charged RNA or other biomolecules hybridize with specific probes attached to the nanomembrane. We have shown extraction and detection of several pathogenic nucleic acids including: microRNA associated with oral cancer, brucellosis, *E.coli* and dengue virus. In addition to nucleic acid detection, we have also demonstrated the ability of the sensor to detect Lipopolysachharide molecules like endotoxin by functionalizing the nanomembrane surface with an antibiotic polymyxin molecule that acts as the probe. We have demonstrated high selectivity (2 base mismatch), large dynamic range (3 decades) and low detection limit (1 fM). The versatility of the platform allows detection of other target biomolecules provided a probe of interest is available and specific.

434794 Selective Biomolecular Detection Using Interfacial Electrokinetic Transduction (IET) Biosensors

Abstract:

Biosensors are analytical devices that combine biological sensing elements with physicochemical transducers to detect and quantify biomolecular interactions. Development of portable sensors capable of rapidly detecting specific biomolecules is of central importance for the diagnosis and treatment of disease, biowarfare/anti-terrorism applications, environmental monitoring, drug discovery, cell biology, cancer research and point-of-care diagnostic testing. The combination of biosensors with microfluidics has potential to fulfill the need for cheap, portable sensors capable of rapid and sensitive detection, as they are amenable to miniaturization, automation and integration. In this work we report the first label-free biosensor for detecting and monitoring bioaffinity interactions in solution at microfluidic liquid interfaces. The approach is based on a novel electrical method that combines continuous flow microfluidics with alternating current (AC) electrokinetics. Bioaffinity binding is initiated at a microfluidic laminar liquid interface. Simultaneously, an AC electrical field is applied to electrokinetically displace the interface across the main flow channel. The magnitude and direction of the interfacial displacement is a function of the applied frequency and the interface electrical properties. We demonstrate that interfacial frequency response is influenced by biomolecular binding. Biosensor detection is based on interfacial electrokinetic transduction (IET); binding changes the electrical properties of the fluid interface, which is electrokinetically transduced, and detected by monitoring the frequency response of the liquid interface. In this manner, the liquid interface serves as both the substrate for specific binding and the transducer for biomolecular recognition. We use IET to electrokinetically detect avidin in solution without labels. The IET biosensor is specific and able to distinguish low concentrations of avidin against a background of bovine serum albumin (BSA). Through this study, we establish a methodology for rapid label-free IET biosensing at electrical liquid interfaces, and demonstrate sensor performance and sensitivity for the detection of biomolecules at low picomolar concentrations. This new electrokinetic approach could provide a low-cost, rapid, and portable biosensing strategy for biomolecular diagnostics.

432463 Sensitive, Multiplexed Electrophoretic miRNA Detection Using Peptide Nucleic Acids in Micelle-Elfse

Abstract:

miRNAs are short, noncoding nucleic acids that are biomarkers for a growing number of disease states, including breast cancer, Alzheimer's disease, and heart disease. Because they are only 20-22 bases in length, and are present in trace quantities, they are difficult to detect by conventional PCR and array methods. Here, we present an electrophoretic detection system that uses synthetic γ-substituted peptide nucleic acids, which bind miRNA with great specificity and stability. Low detection limits are achieved using an ultrabright probe in the form of a DNA oligomer saturated with intercalating dye. The micelle-ELFSE method has been extended to kilobase-length DNA to provide ultrabright fluorophores for sub-attomole detection of miRNA in less than 3 minutes. The use of surfactant running buffers, rather than gels, makes the method compatible with high amounts of serum (10-100 mg/ml) with minimal peak distortion. Finally, the use of gamma-substituted peptide nucleic acid probes gives great binding stability and specificity, allowing for miRNA detection in a sandwich format. The method can easily distinguish members of the closely related let-7 series. We will discuss efforts to concentrate miRNA by ITP-based sample stacking as well as theoretical limits of peak capacity for multiplexed detection in capillary and microchip electrophoresis. We will also discuss applications in early-stage cancer screening and other medical applications of this rapid, selective miRNA detection scheme.

AC Electrokinetics of conducting micro- and nano-particles

by Antonio Ramos

Collaborators:
Pablo García-Sánchez[1]
Juan J. Arcenegui[1]
Hywel Morgan[2]
[1]Univ. Seville, [2]Univ. Southampton

AES 2015 Annual Meeting

Nov 9–11 2015, Salt Lake City, Utah, US

MEC Project FIS2014-54539-P

Outline

- Introduction: Metal particles in electrolyte
- DEP and ROT of microspheres
- EOr and ROT of nanowires
- Electrokinetic interactions between nanowires

Outline

- Introduction: Metal particles in electrolyte
- DEP and ROT of microspheres
- EOr and ROT of nanowires
- Electrokinetic interactions between nanowires

A metal particle in an electrolyte is subjected to an electric field...

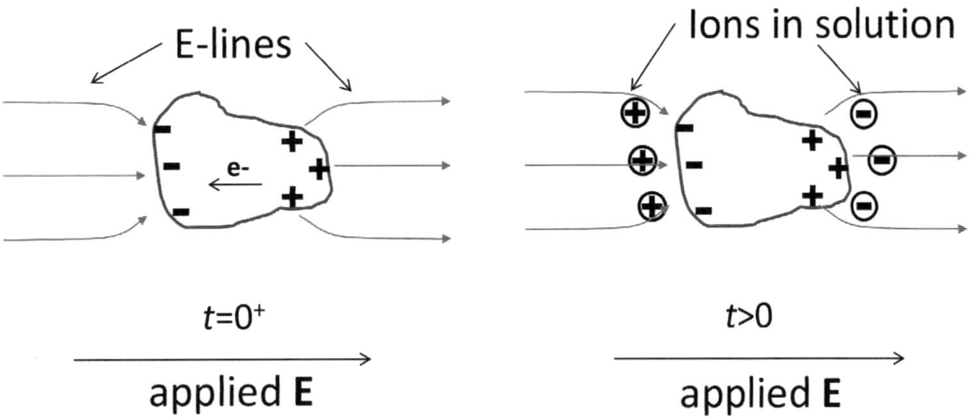

There are two distinctive limits in the steady state

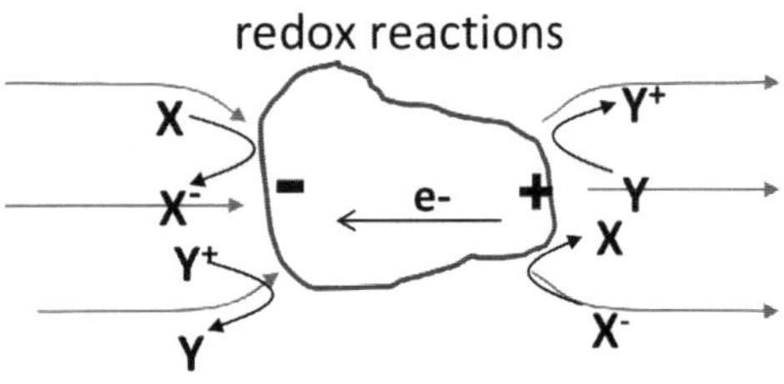

$t=\infty$

applied **E** >>

$$E\ell > V_{\text{redox}}$$

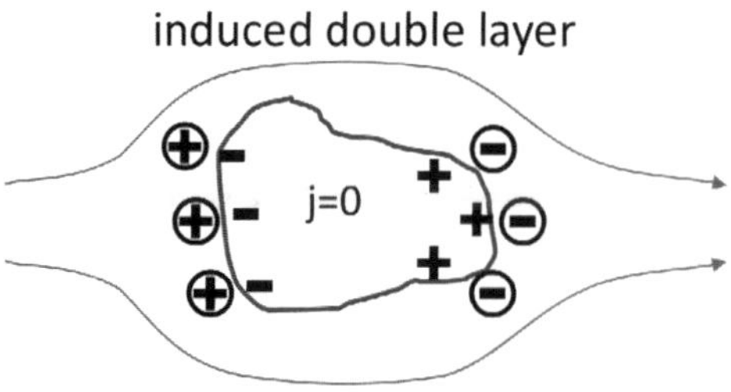

$t=\infty$

applied **E** <<

$$E\ell < V_{\text{redox}}$$

for water electrolysis is 1.23 volts

Example $E \sim 10^4 - 10^5$ V/m, $\ell \sim 10^{-5}$ m, $E\ell$ ~0.1-1 V

Particle polarization changes from low to high frequencies in AC fields

applied **E** <<

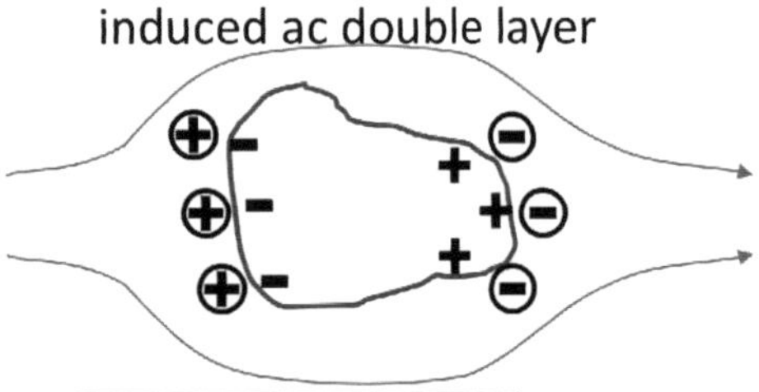

$\omega << 1/RC_{\text{DL}}$

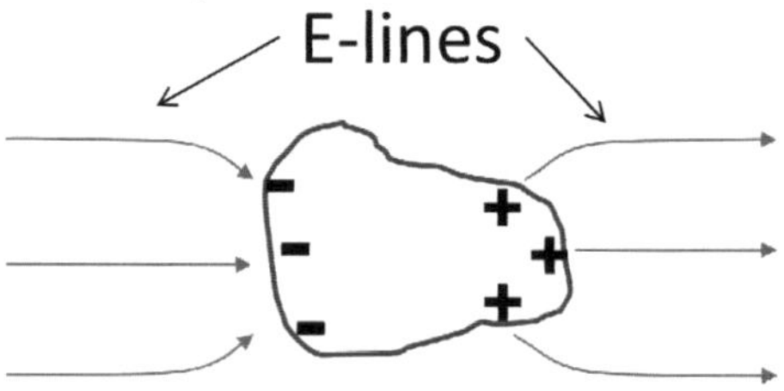

$\omega >> 1/RC_{\text{DL}}$

RC_{DL} time ~$\ell\,\varepsilon/\sigma\lambda_{\text{D}}$

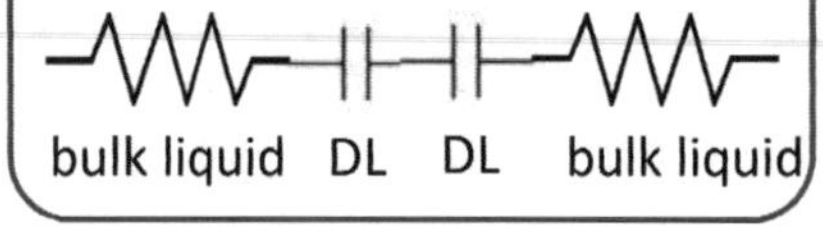

Metal colloids?

To be stable we expect them higly charged. Intrinsic double layer may contribute to surface conductance -> high Dukhin number

$Du = \sigma_s/(\sigma a)$

(Lyklema's book)

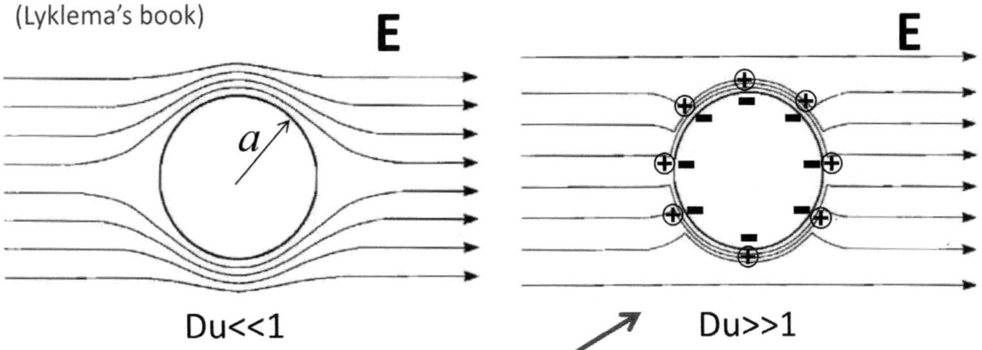

We could expect induced positive dipole for all frequencies

$a \ll 1\ \mu m$

Mathematical formulation for electrical potential (small voltage and thin-double-layer approximations)

$E \cdot l \sim \Delta\phi < k_B T/e = 25$ mV

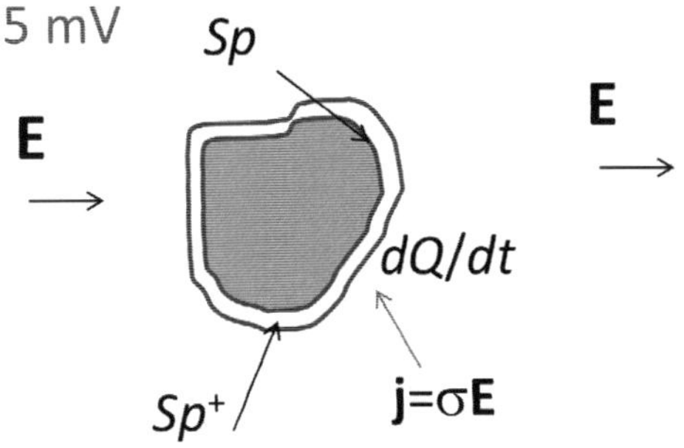

Laplace's equation in bulk

$$\nabla^2 \phi = 0$$

B.C. Charging of DL

$$\sigma \frac{\partial \phi}{\partial n} = C_{DL} \frac{\partial \Delta\phi}{\partial t}$$

$$\Delta\phi = \phi]_{S_p^+} - V$$

Particle motion is due to:
a) electrical forces on particle charges and
b) electrical forces on free charge in liquid transmitted to particle by hydrodynamic stresses

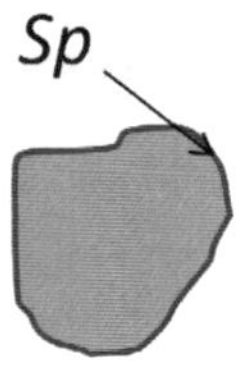

outside *Sp*

$$\eta\nabla^2\vec{u} - \nabla p + \rho_q\vec{E} = 0 \leftrightarrow \nabla\cdot\left(\overleftrightarrow{\Sigma}_H + \overleftrightarrow{\Sigma}_E\right) = 0$$

b.c. $\vec{u} = \vec{U} + \vec{\Omega}\times\vec{r}$ on *Sp*

no inertia, low Reynolds number

$$\int_{Sp}\left(\overleftrightarrow{\Sigma}_H + \overleftrightarrow{\Sigma}_E\right)\cdot d\vec{S} = 0$$ equilibrium of forces

$$\int_{Sp}\vec{r}\times\left(\overleftrightarrow{\Sigma}_H + \overleftrightarrow{\Sigma}_E\right)\cdot d\vec{S} = 0$$ equilibrium of torques

equilibrium of forces and torques determine **U** and **Ω**

Equations for particle motion due to electric field for thin double layers

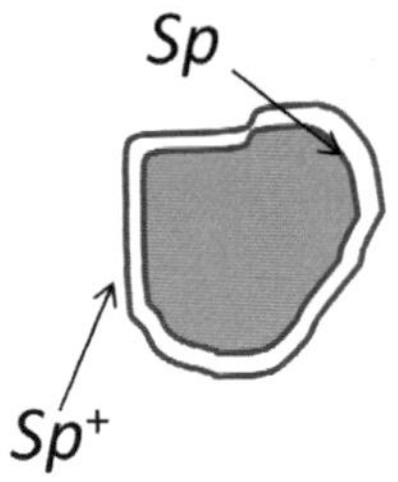

$$\eta\nabla^2\vec{u} - \nabla p = 0$$ outside *Sp*+

b.c. $\vec{u}|_{Sp+} = \vec{U} + \vec{\Omega}\times\vec{r} + \vec{u}_{ICEO}$ on *Sp*+

$$\Delta\phi = \phi]_{S_p^+} - V \qquad \vec{u}_{HS} = \frac{\varepsilon}{\eta}\Delta\phi\vec{E}_s$$

Helmholtz-Smoluchowski slip velocity

Equations for particle motion due to electric field

$$\vec{u}|_{Sp+} = \vec{U} + \vec{\Omega} \times \vec{r} + \vec{u}_{ICEO}$$

superposition of solutions:
$u=u_U+u_\Omega+u_{ICEO}$

$$\int_{S_\infty} (\overleftrightarrow{\Sigma}_H + \overleftrightarrow{\Sigma}_E) \cdot d\vec{S} = 0$$

$$\rightarrow \quad -\eta \overleftrightarrow{K} \cdot \vec{U} - \cancel{\eta \overleftrightarrow{C} \cdot \vec{\Omega}} + \int_{S_\infty} \overleftrightarrow{\Sigma}_H^{ICEO} \cdot d\vec{S} + \vec{F}_E = 0$$

$$\int_{S_\infty} \vec{r} \times (\overleftrightarrow{\Sigma}_H + \overleftrightarrow{\Sigma}_E) \cdot d\vec{S} = 0$$

$$\rightarrow \quad -\eta \overleftrightarrow{K}_r \cdot \vec{\Omega} - \cancel{\eta \overleftrightarrow{C}^t \cdot \vec{U}} + \int_{S_\infty} r \times \overleftrightarrow{\Sigma}_H^{ICEO} \cdot d\vec{S} + \vec{T}_E = 0$$

For particles without screw-like properties coupling tensor *C* vanishes if origin is properly chosen

Equations for particle motion due to electric field

for particles without screwlike properties

p is dipole of (particle + DL)

$$\eta \overleftrightarrow{K} \cdot \vec{U} = \int_{S_\infty} \langle \overleftrightarrow{\Sigma}_H^{ICEO} \rangle \cdot d\vec{S} + \langle (\vec{p} \cdot \nabla)\vec{E} \rangle$$

$$\vec{U} = \vec{U}_{ICEO} + \vec{U}_{DEP}$$

time average for AC fields

$$\eta \overleftrightarrow{K}_r \cdot \vec{\Omega} = \int_{S_\infty} r \times \langle \overleftrightarrow{\Sigma}_H^{ICEO} \rangle \cdot d\vec{S} + \langle \vec{p} \times \vec{E} \rangle$$

$$\vec{\Omega} = \vec{\Omega}_{ICEO} + \vec{\Omega}_{DEP}$$

Example: metal sphere subjected to electric field (thin-double-layer approximation)

$$\vec{U} = \vec{U}_{ICEO} + \frac{\langle(\vec{p}\cdot\nabla)\vec{E}\rangle}{6\pi a\eta} \qquad \vec{\Omega} = \vec{\Omega}_{\mathrm{ICEO}} + \frac{\langle\vec{p}\times\vec{E}\rangle}{8\pi a^3\eta}$$

(Fair & Anderson, 1989)

$$\vec{U}_{ICEO} = -\frac{\int \vec{u}_{HS}\, dS}{4\pi a^2} \qquad \vec{\Omega}_{ICEO} = -\frac{3}{8\pi a^3}\int \vec{n}\times\vec{u}_{HS}\, dS$$

$$\vec{u}_{HS} = \frac{\varepsilon}{\eta}\langle\Delta\phi\vec{E}_s\rangle$$

Helmholtz-Smoluchowski slip velocity

$$\Delta\phi = \phi]_{S_p^+} - V$$

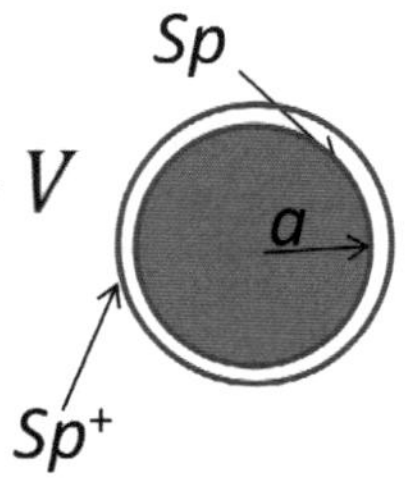

$\mathbf{U}_{\mathrm{ICEO}}$ and $\mathbf{\Omega}_{\mathrm{ICEO}}$ tend to zero for frequencies much higher than $1/RC_{DL}$

Particle motions due to AC electric field

Translational motion

Homogeneoud field + asymmetric particle

(Squires-Bazant 2006)

(Gangwal et al 2008)

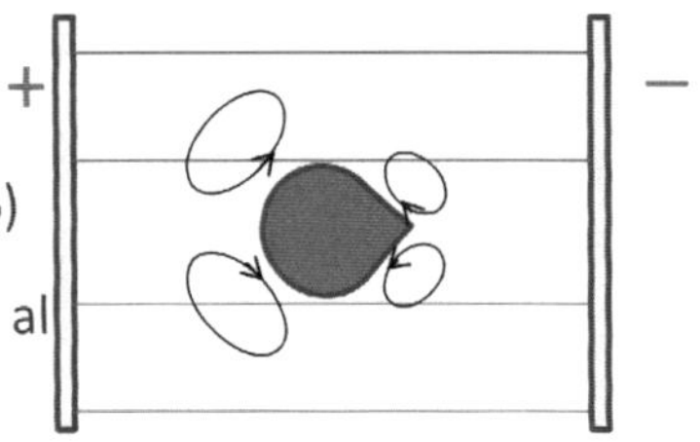

$$\vec{U} = \vec{U}_{ICEO}$$

IC EO flow around particle=
IC Electrophoresis

Inhomogeneous field

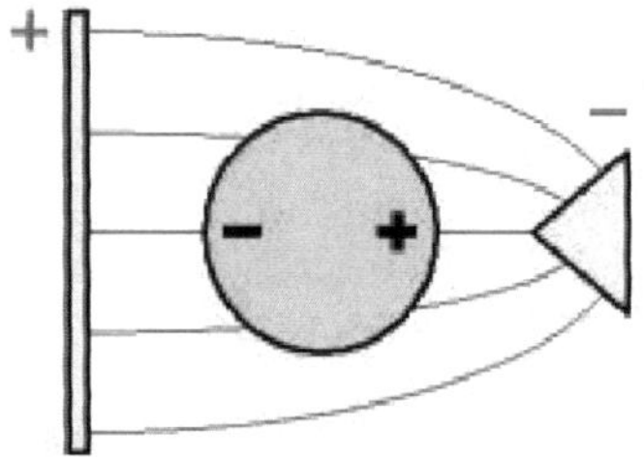

$$\vec{U} = \vec{U}_{ICEO} + \vec{U}_{DEP}$$

Dielectrophoresis
+ IC Electrophoresis

U_{ICEO} tends to zero for frequencies much higher than $1/RC_{DL}$

Particle motions due to AC electric field

Rotational motion

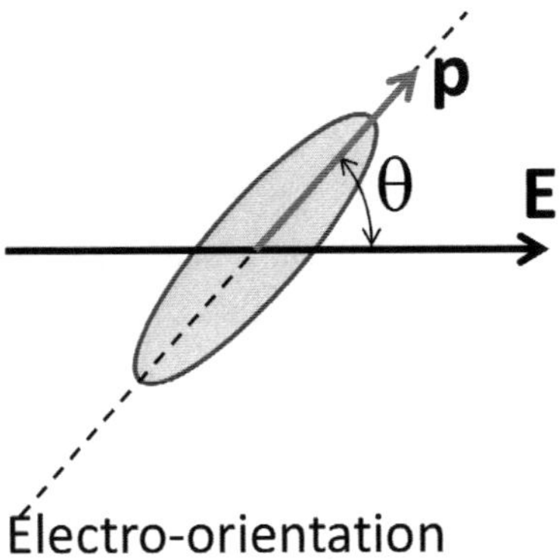

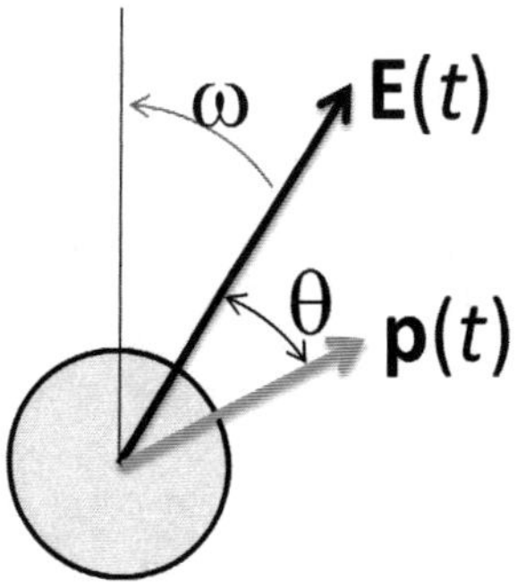

Electro-orientation

Electro-rotation

$$\vec{\Omega} = \vec{\Omega}_{ICEO} + \vec{\Omega}_{DEP}$$

Ω_{ICEO} tends to zero for frequencies much higher than $1/RC_{\text{DL}}$

Outline

- Introduction: Metal particles in electrolyte
- DEP and ROT of microspheres
- EOr and ROT of nanowires
- Electrokinetic interactions between nanowires

In Dielectrophoresis, force on induced dipole is proportional to real part of polarizability

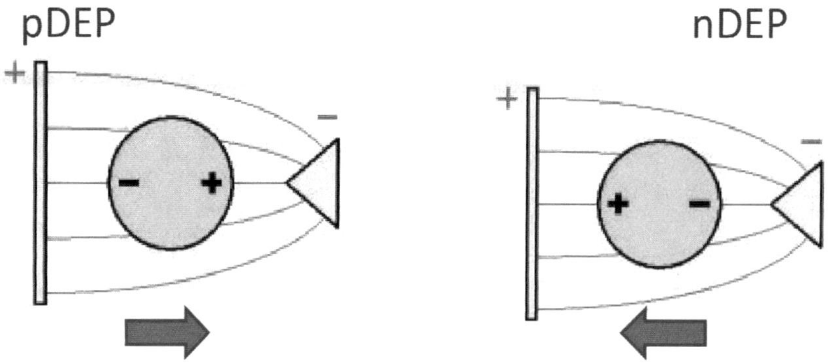

$$\vec{F}_{DEP} = \langle (\vec{p} \cdot \nabla)\vec{E} \rangle = 2\pi a^3 \varepsilon_m \mathrm{Re}[K(\omega)] \nabla E_{rms}^2$$

In Electrorotation, torque on induced dipole is proportional to imaginary part of polarizability

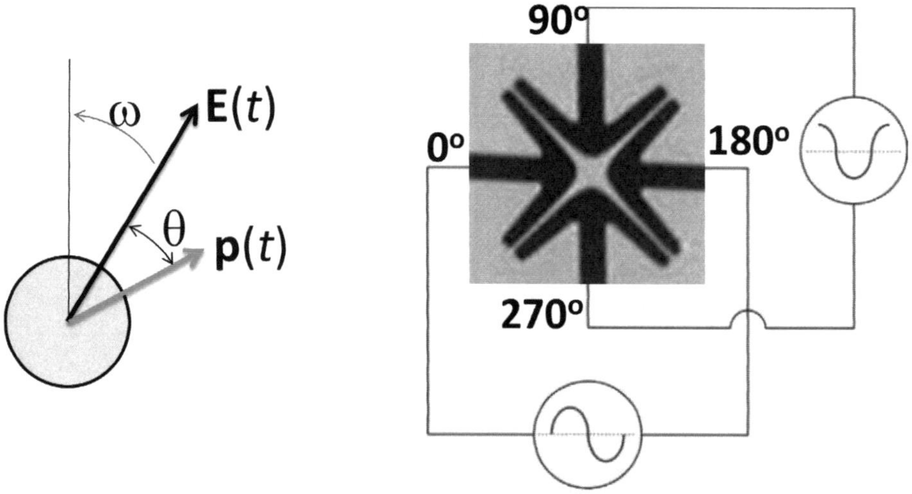

$$\vec{T}_{ROT} = \langle \vec{p} \times \vec{E} \rangle = -4\pi a^3 \varepsilon_m \mathrm{Im}[K(\omega)] E_0^2 \hat{z}$$

Dipole expression is obtained from Laplace's equation

general potential solution

$$\phi = -E_0 r\cos\theta + \frac{p}{4\pi\varepsilon r^2}\cos\theta$$

$$\nabla^2\phi = 0$$

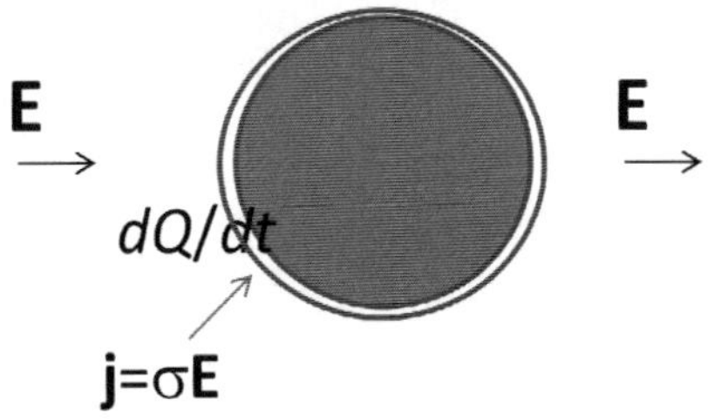

boundary condition

$$\sigma\frac{\partial\phi}{\partial r}\bigg]_{a^+} = i\omega C_{DL}(\phi(a^+) - V)$$

dipole expression $\vec{p} = 4\pi\varepsilon a^3 K(\omega)\vec{E}$

Claussius-Mossotti $K(\omega) = \dfrac{i\omega\tau - \frac{1}{2}}{i\omega\tau + 1}$ $\quad \tau = \dfrac{C_{DL}a}{2\sigma}$ RC time

Electric field lines around a metal sphere at low and high frequencies: from insulator to conductor

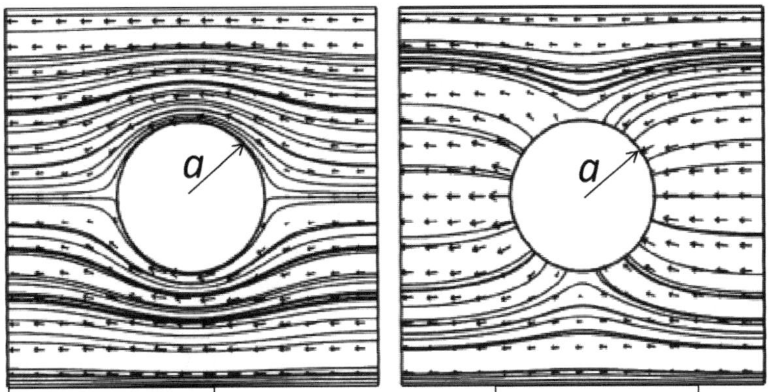

$\omega \ll 1/\tau$ $\qquad$ $\omega \gg 1/\tau$

$$\tau = C_{DL}a/2\sigma \sim \varepsilon a/2\sigma\lambda_D$$

Debye-Huckel approx

Force and torque expressions

$$K(\omega) = \frac{i\omega\tau - \frac{1}{2}}{i\omega\tau + 1}$$

$$\tau = \frac{C_{DL}\, a}{2\sigma}$$

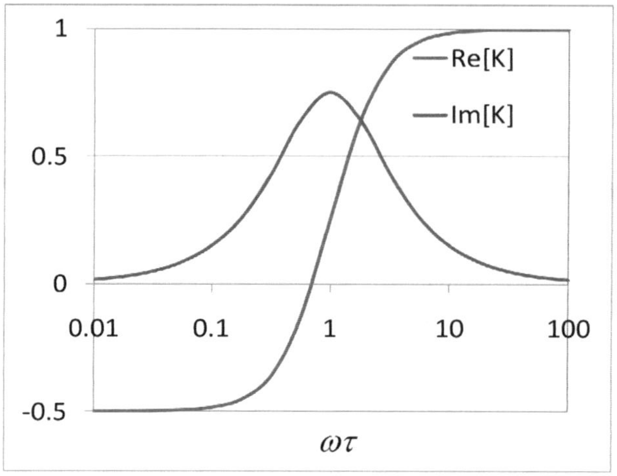

$\vec{F}_{DEP} = 2\pi a^3 \varepsilon_m \mathrm{Re}[K(\omega)] \nabla E_{rms}^2$ nDEP to pDEP increasing frequency

$\vec{T}_{ROT} = -4\pi a^3 \varepsilon_m \mathrm{Im}[K(\omega)] E_0^2 \hat{z}$ ROT is counter-field

To this expressions we must add ICEO effect at low frequencies

Translation

Motion at low frequencies is a combination of DEP and ICEP. It was called dipolophoresis by Shilov.

Perfectly polarizable metal sphere should not move in DC fields. ICEP+DEP=0
(Shilov-Simonova, 1981)
(Squires-Bazant, 2006)
(Miloh, 2009)

Rotation

ROT is not affected because expected ICEO flow is not rotating:

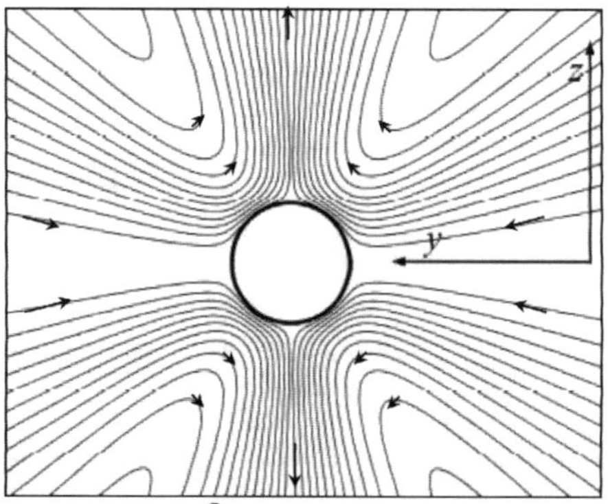

$$\vec{\Omega}_{ICEO} = -\frac{3}{8\pi a^3}\int \vec{n} \times \vec{u}_{HS}\, dS = 0$$

for thin DL

Experiments with gold-coated polystyrene microspheres

(Acknowledgements: Y. Ren and D. Morganti)

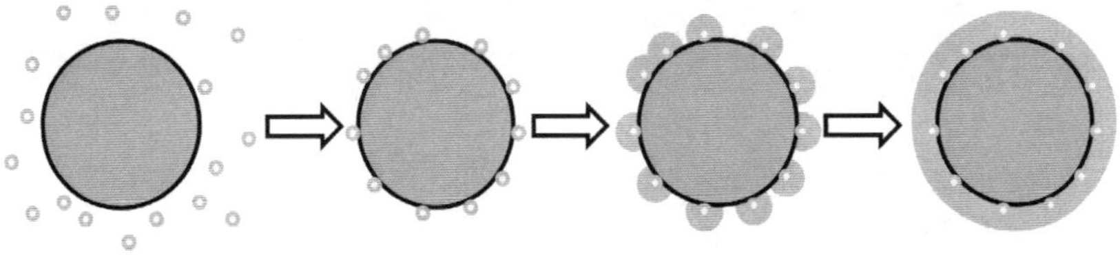

polystyrene beads (10, 25, 40μm diameter) were coated with colloidal gold (15nm)

particle is almost neutrally buoyant in water

DEP of Gold-coated polystyrene microspheres

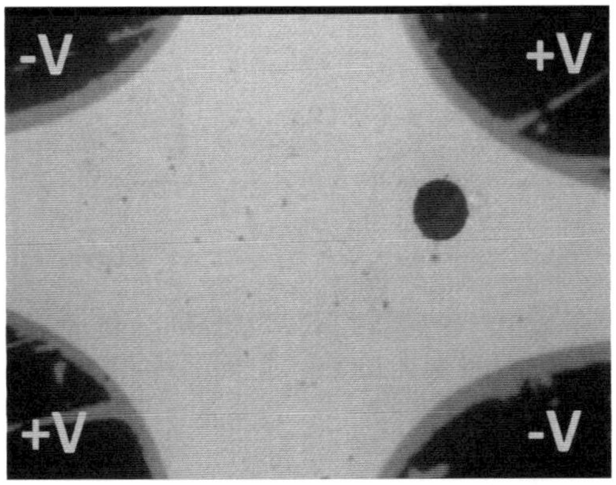

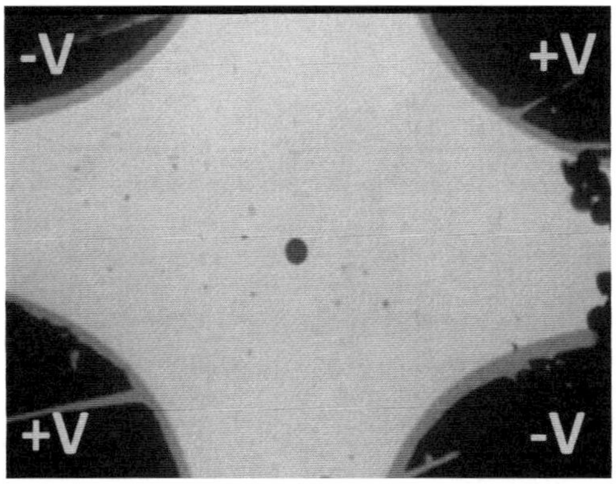

Negative DEP
f = 100 Hz, σ=3.6 mS/m

Positive DEP
f = 10MHz, σ=0.7 mS/m

negligible U_{ICEO} for these metal-coated spheres
Probably: surface roughness and compact layer

ROT of Gold-coated polystyrene microspheres is counter-field

E rotation

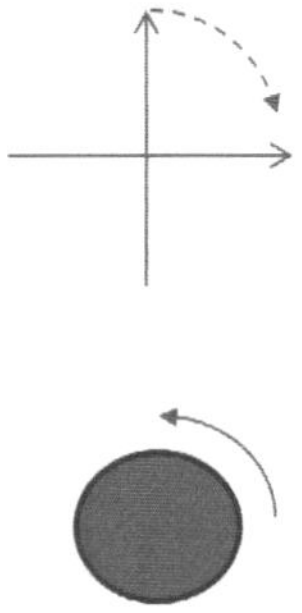

particle rotation

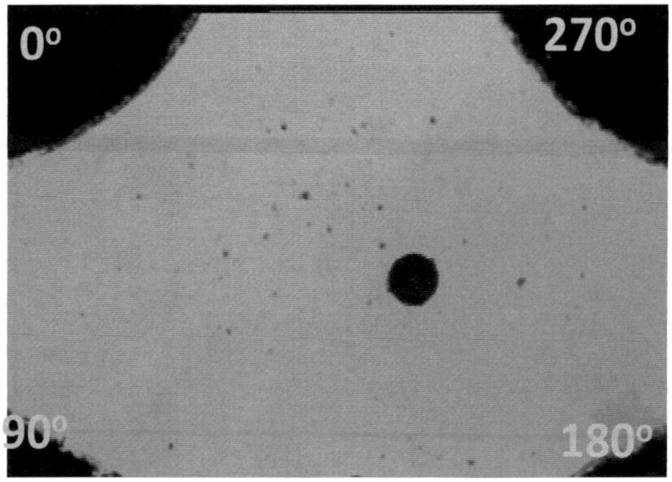

V= 6 V_{pp}
f = 200 Hz
σ=0.7 mS/m

ROT peaks (left) correspond to maximum variation of DEP (right)

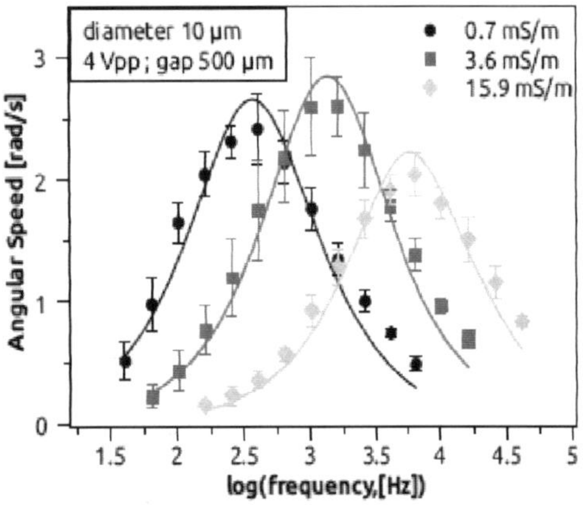

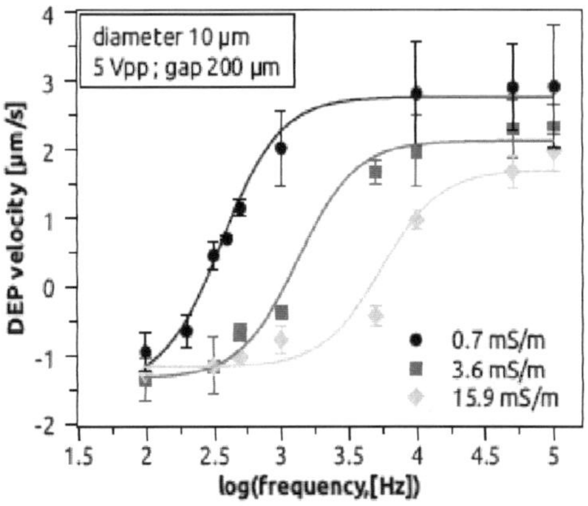

(García-Sánchez et al, Langmuir 2012)

Experiments with Titanium microspheres

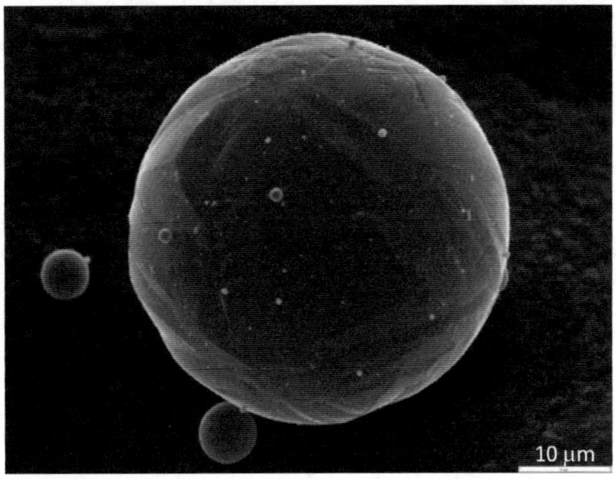

Ti spheres with a range of radii from 1 to 25 μm

Ti spheres are around 70 times heavier than polystyrene spheres. Very difficult to do DEP

ROT of Titanium microspheres

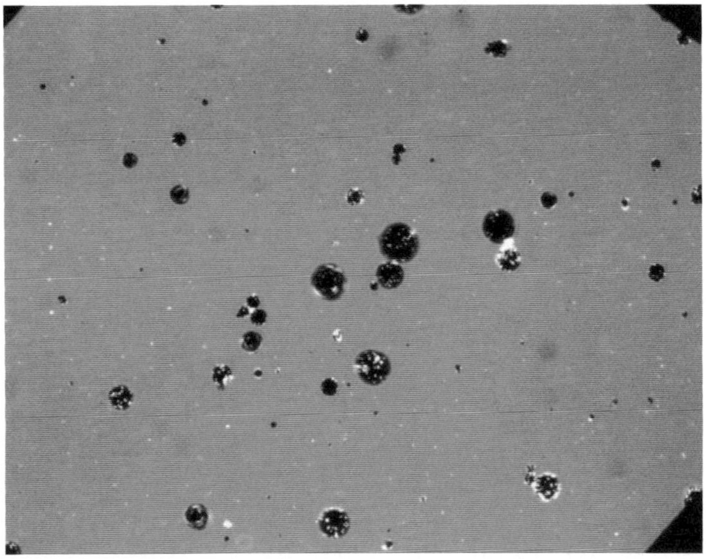

f=20 kHz, V=8 V_{pp} , σ=15.9 mS/m

ROT of Titanium microspheres

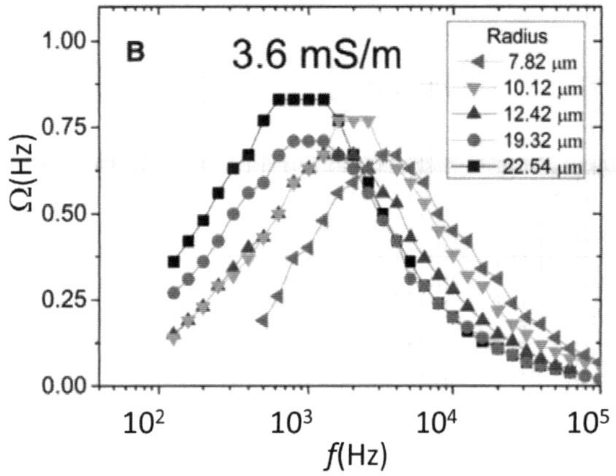

Rotation speed versus signal frequency

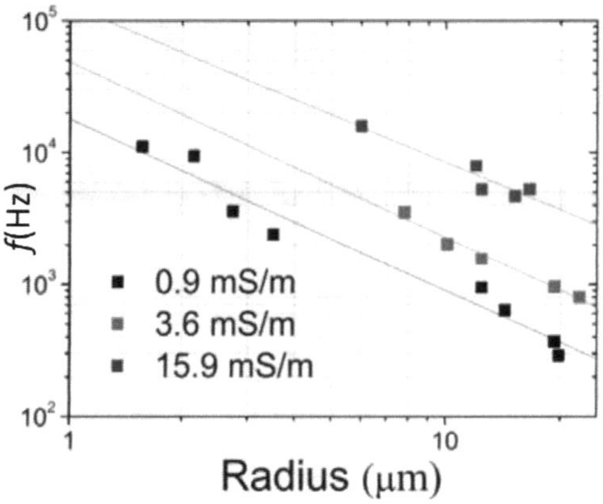

Max signal frequency versus radius

(Arcenegui et al, Electrophoresis 2013)

Viscous torque near wall balances ROT torque

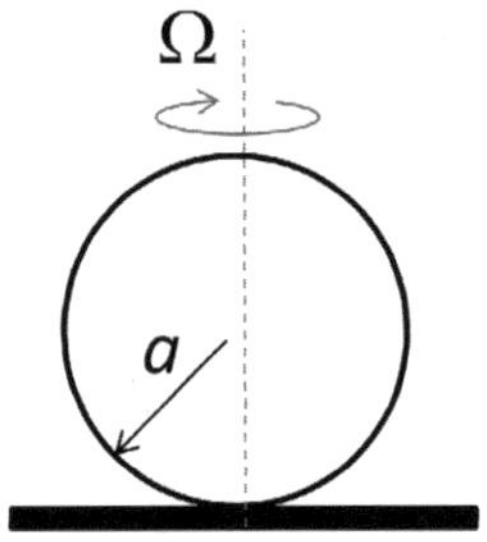

$$\vec{T}_{vis} = 8\pi\eta a^3 \Omega \zeta(3) \hat{z}$$ (Jeffery, 1915)

$$\zeta(3) = 1.20206\ldots$$

$$\vec{T}_{ROT} = \vec{T}_{vis} \rightarrow \Omega = -\frac{\varepsilon_m \mathrm{Im}[K(\omega)] E_0^2}{2\zeta(3)\,\eta}$$

$$\Omega_{\max} = -0.312\,\frac{\varepsilon_m E_0^2}{\eta}$$

most measured maximum rotational speeds within 20% error!
for both gold-coated and Ti spheres

Viscous force near wall balances DEP force

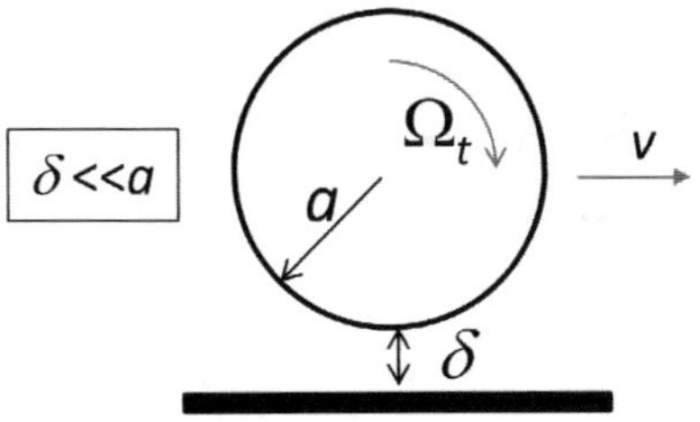

$$F_{vis} = \frac{4\pi\eta a}{5}\ln(\delta/a)[4v - a\Omega_t]$$

(Goldman, Cox, Brenner, 1967)

Exp observation: our **gold-coated spheres** do not rotate, Ω_t=0

$\mathbf{F}_{DEP}+\mathbf{F}_{visc}=0$ ⟹ $$\vec{v} = \frac{5\, a^2\varepsilon_m \mathrm{Re}[K(\omega)]\nabla E_0^2}{16\eta \ln(a/\delta)}$$

velocity values on the order of experimental ones for δ=15 nm (typical size of asperities!!!)

Outline

- Introduction: Metal particles in electrolyte
- DEP and ROT of microspheres
- EOr and ROT of nanowires
- Electrokinetic interactions between nanowires

Electro-orientation

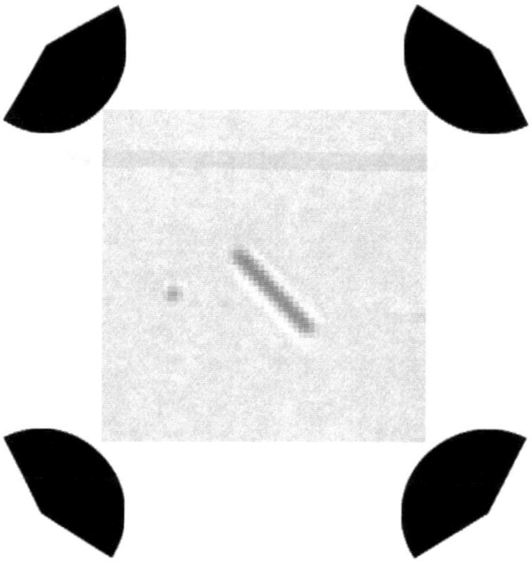

silver nanowires
length ~ 6 microns, diameter ~ 0.3 microns

Electro-rotation

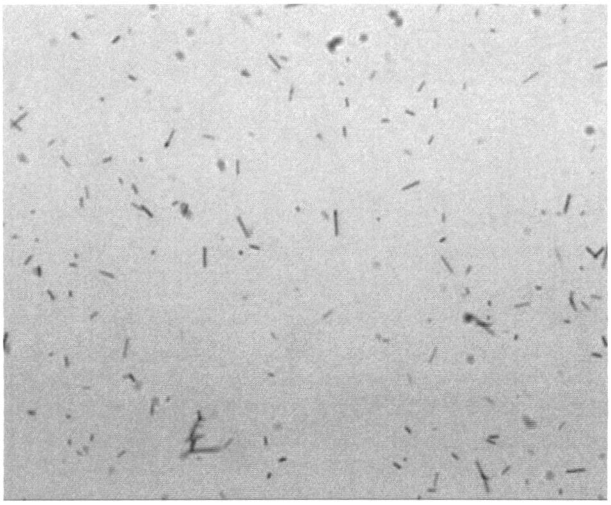

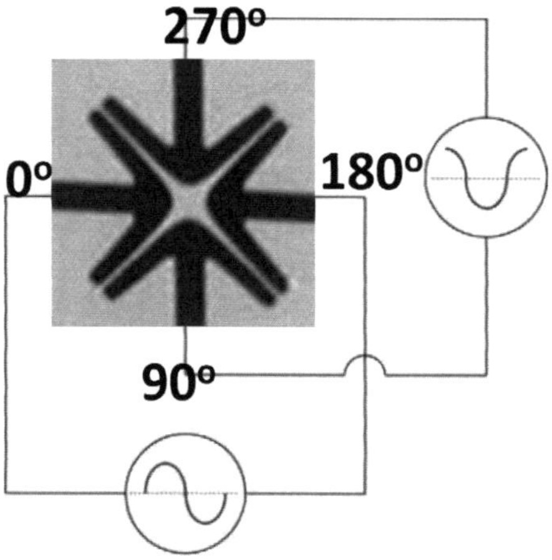

ROT is counterfield

Dipole theory

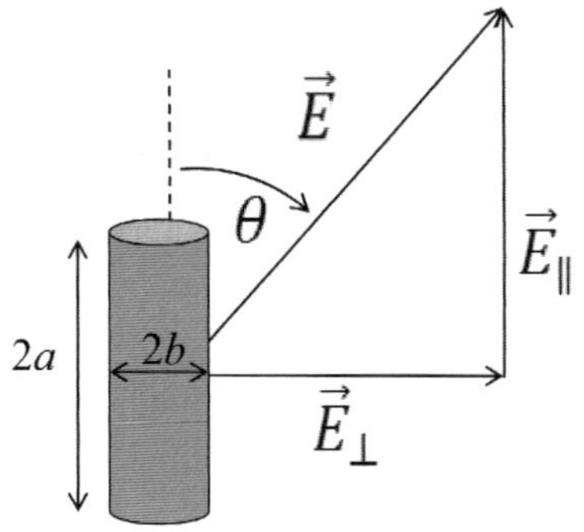

slender cylinder

$$\vec{p} = \alpha_\parallel \vec{E}_\parallel + \alpha_\perp \vec{E}_\perp \approx \alpha_\parallel \vec{E}_\parallel$$

$$\langle \vec{T}_E \rangle = \frac{1}{2}\mathrm{Re}[\vec{p} \times \vec{E}^*] \approx \vec{e}_z \frac{1}{2}\mathrm{Re}[\alpha_\parallel E_\parallel E_\perp^*]$$

$$T_E = T_v = \gamma\dot{\theta}$$

Electro-orientation

$$\langle T_E \rangle = \frac{1}{2}E_0^2\,\mathrm{Re}[\alpha_\parallel]\cos\theta\sin\theta$$

Electro-rotation

$$\langle T_E \rangle = -\frac{1}{2}E_0^2\,\mathrm{Im}[\alpha_\parallel]$$

Induced dipole of metal nanowire in electrolyte versus frequency: from insulator to conductor

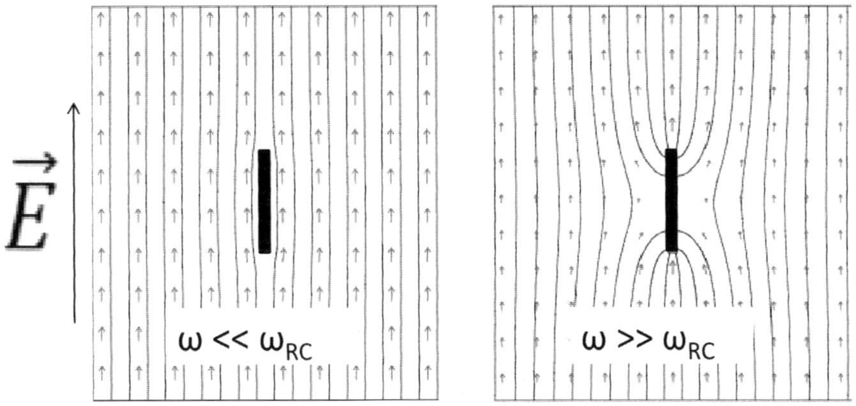

$$\nabla^2\phi = 0 \qquad \text{b.c.} \quad \sigma\nabla\phi \cdot \vec{n} = i\omega q_s = i\omega C_{DL}\phi$$

characteristic angular frequency $\omega_{RC} = \dfrac{\sigma}{aC_{DL}} f(b/a)$

Polarizability can be computed numerically

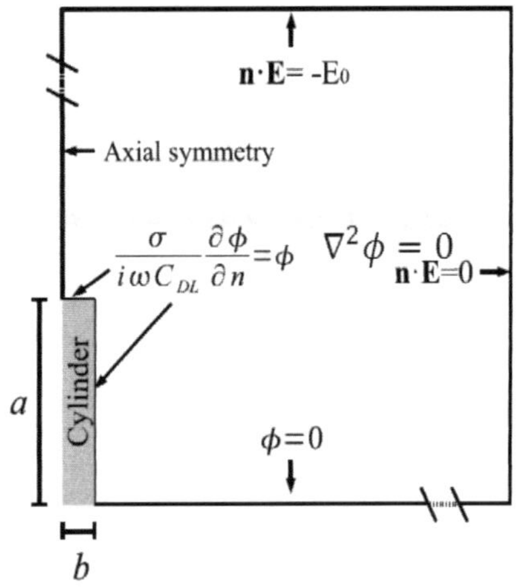

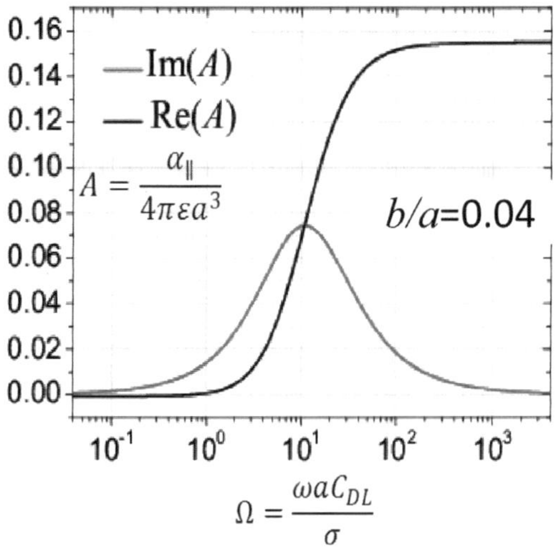

$$\dot{\theta} = \frac{T_E}{\gamma}$$

Experimental results

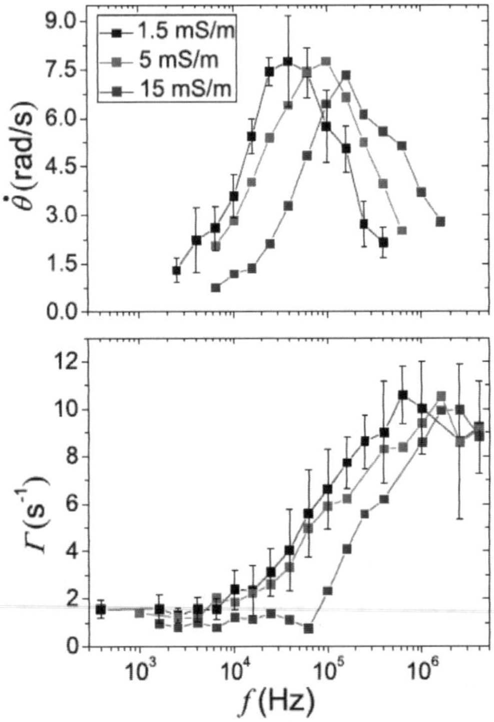

Electro-rotation

$$\dot{\theta} = \dot{\theta}(\omega)$$

Electro-orientation

$$\dot{\theta} = \Gamma(\omega)\sin(2\theta)$$

orientation rate

$$\tan\theta = \tan\theta_0 e^{-2\Gamma t}$$

C_{DL} is obtained from maximum in ROT and is close to Debye-Huckel capacitance

$$\Omega_{\max} = \frac{\omega a C_{DL}}{\sigma} \approx 10$$

TABLE I. Values for the electrical double layer capacitance obtained from experiments (C_{DL}), Debye-Hückel theoretical values for the Debye layer (C_{DH}), and best-fit values for Debye and compact layers in series (C_{S+DH}).

Conductivity (mS/m)	C_{DL} (F m^{-2})	C_{DH} (F m^{-2})	C_{S+DH} (F m^{-2})
1.5	1.9×10^{-2}	2.3×10^{-2}	1.9×10^{-2}
5	3.5×10^{-2}	4.2×10^{-2}	3.0×10^{-2}
15	4.3×10^{-2}	7.3×10^{-2}	4.4×10^{-2}

Experimental results compared to dipole theory

(Arcenegui et al, PRE 2013)

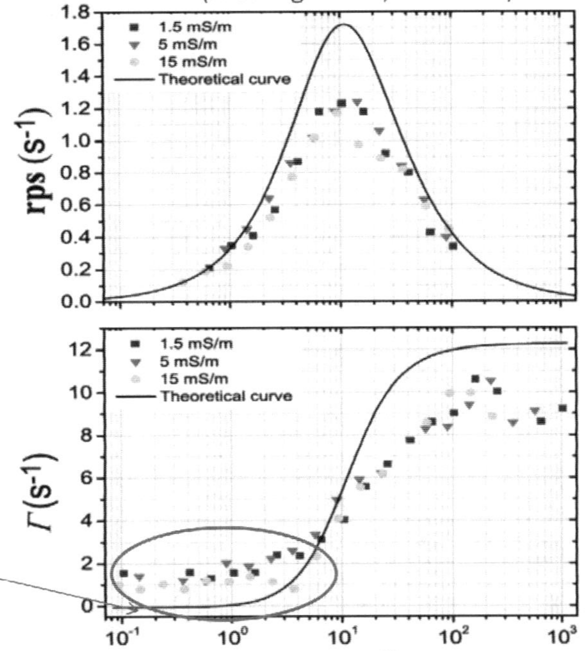

$$\dot{\theta} = \frac{T_E}{\gamma}$$

This cannot be explained by dipole theory

$$\Omega = \frac{\omega a C_{DL}}{\sigma}$$

Electro-rotation ~ Im(A)

$$\dot{\theta} = \frac{4\pi\varepsilon a^3 E_0^2}{2\gamma} \mathrm{Im}\,[A]$$

Electro-orientation ~ Re(A)

$$\Gamma = \frac{4\pi\varepsilon a^3 E_0^2}{2\gamma} \mathrm{Re}\,[A]$$

At low frequencies, there is induced charge electroosmotic flow

(Squires Bazant JFM 2004)

(Saintillan JFM 2006)

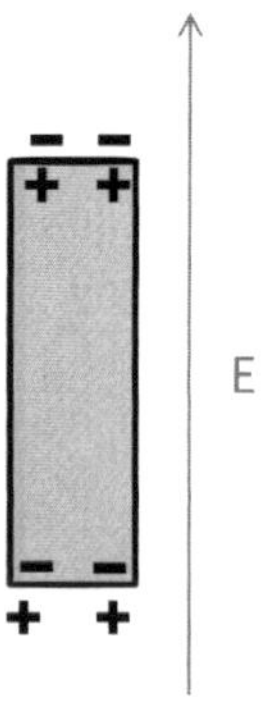

low frequency and perfectly polarizable DL is fully charged

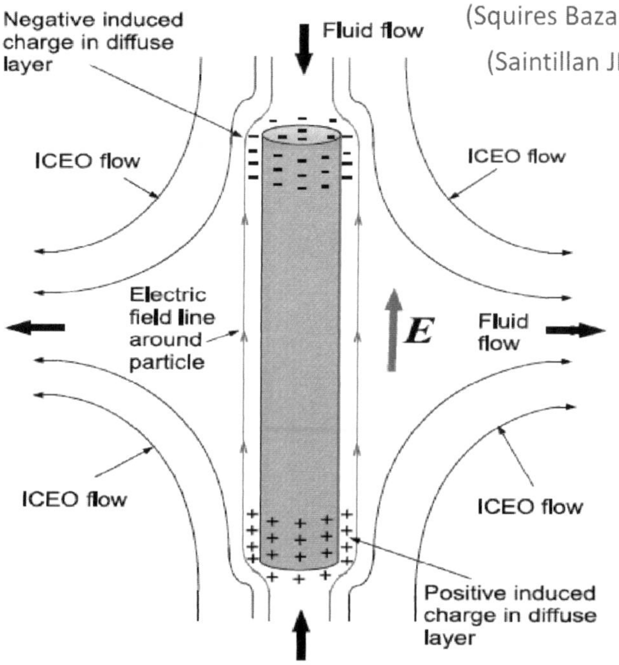

Rotation is due to the cross-term induced charge due to E_z multiplied by E_x

Similar to particle in EP

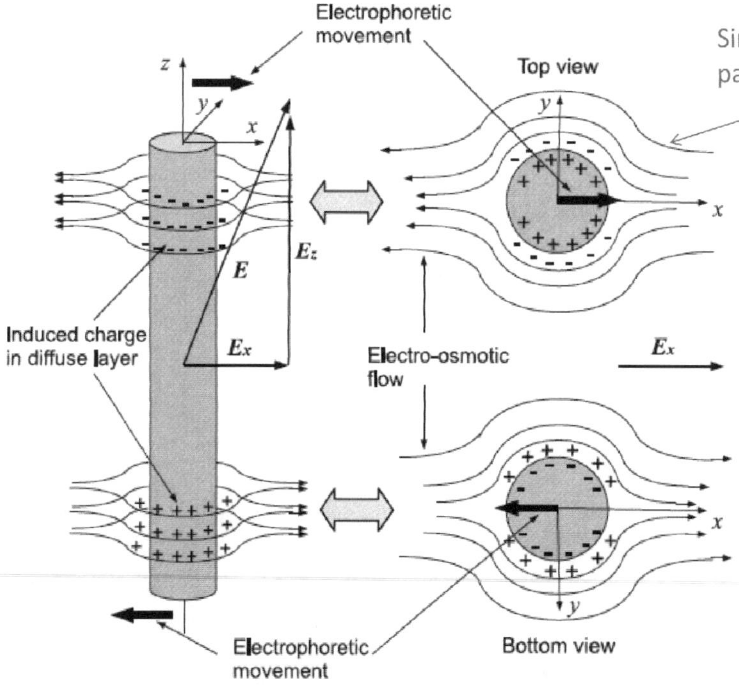

We obtained the angular velocity of nanowire from slender-body and thin layer approximations

$$\vec{\Omega} = -\frac{3}{2a^3}\int_{-a}^{a} s\vec{e}_{\parallel} \times \tilde{u}_p(s)ds + \frac{3\ln(a/b)}{8\pi\eta a^3}\vec{T}_E$$

(Saintillan JFM 2006)

where $$\tilde{u}_p(s) = \frac{1}{2\pi}\int_0^{2\pi} \vec{u}_p(s,\phi)\, d\phi$$

Helmholtz Smoluchowski velocity (given by electrical problem) as well as the electrical torque T_E

Electrical potential is superposition of: potential due to parallel field (a) and perpendicular field (b)

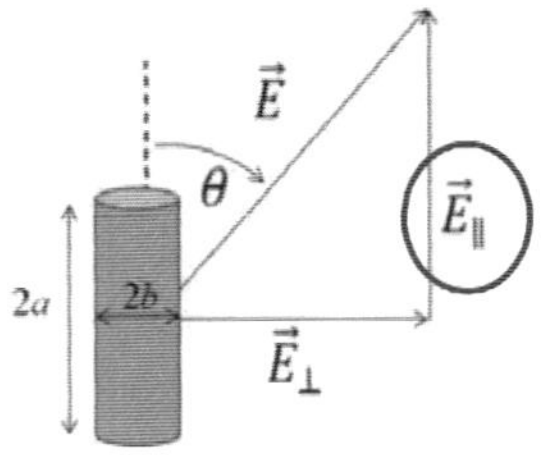

Potential for parallel E; slender-body approximation

$$\Phi_a = -E_{0a}z + \frac{1}{4\pi\varepsilon}\int_{-a}^{a} \frac{\lambda(z')}{\sqrt{(z-z')^2+\rho^2}}dz'$$

from b.c. linear charge λ is obtained $$i\omega C_{\mathrm{DL}}\Phi_a|_{\rho=b} = \sigma \left.\frac{\partial \Phi_a}{\partial \rho}\right|_{\rho=b}$$

slender-body approximation of integrals $$\Phi_a(z,b) \sim -E_{0a}z + \frac{\lambda(z)}{2\pi\varepsilon}\ln(a/b)$$

$$\left.\frac{\partial \Phi_a}{\partial \rho}\right|_b = -\frac{1}{4\pi\varepsilon}\int_{-a}^{a} \frac{\lambda(z')b}{((z-z')^2+b^2)^{3/2}}dz' \sim -\frac{\lambda(z)}{2\pi\varepsilon b}$$

The potential for perpendicular E (problem b) is that of an infinite cylinder

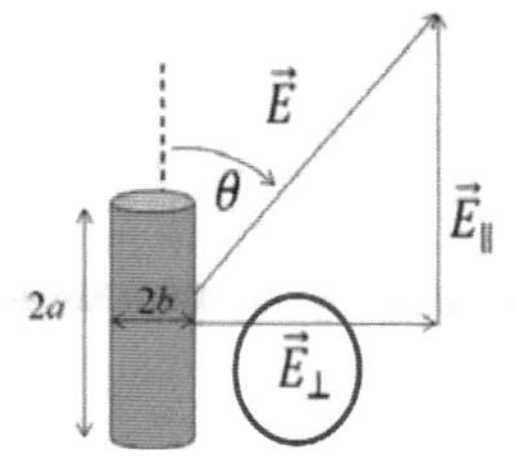

$$\Phi_b = -E_{0b}\rho\cos\phi + \left(\frac{i\Omega_b - 1}{i\Omega_b + 1}\right)\frac{E_{0b}b^2\cos\phi}{\rho}$$

Angular velocity due to induced dipole

EOr $$\dot{\theta} = \frac{\varepsilon E_0^2}{4\eta}\frac{\Omega_b^2\ln^2(1/\beta)}{1+\Omega_b^2\ln^2(1/\beta)}\cos\theta\sin\theta$$

ROT $$\dot{\theta} = -\frac{\varepsilon E_0^2}{4\eta}\frac{\Omega_b\ln(1/\beta)}{1+\Omega_b^2\ln^2(1/\beta)}$$ counter-field rotation

Angular velocity due to induced HS velocity

EOr $$\dot{\theta} = \frac{\varepsilon}{2\eta}\frac{E_0^2\cos\theta\sin\theta}{1+\Omega_b^2\ln^2(1/\beta)}$$ at zero frequency it coincides with (Saintillan JFM 2006, Yariv JFM 2008)

ROT $$\dot{\theta} = \frac{\varepsilon E_0^2}{2\eta}\frac{\Omega_b\ln(1/\beta)}{1+\Omega_b^2\ln^2(1/\beta)}$$ cofield rotation

Angular velocity

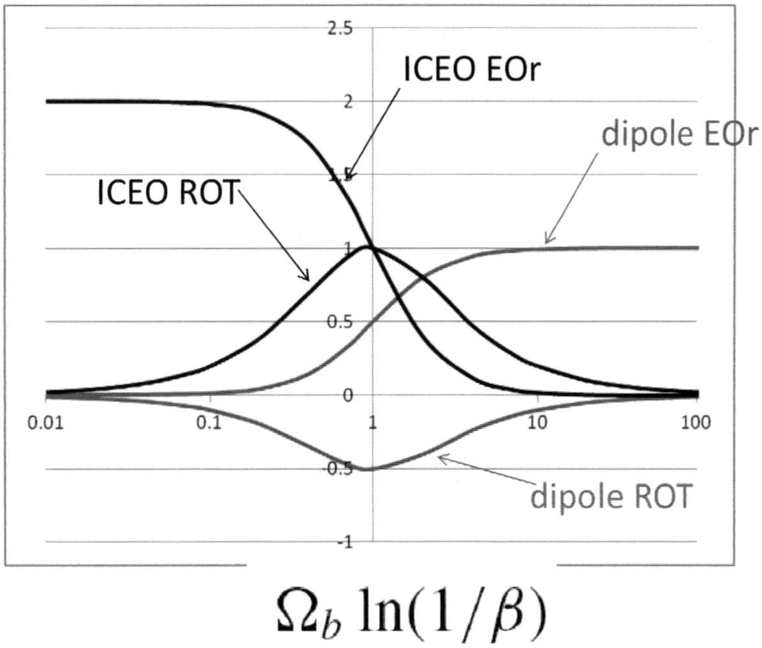

Comparison with experiments

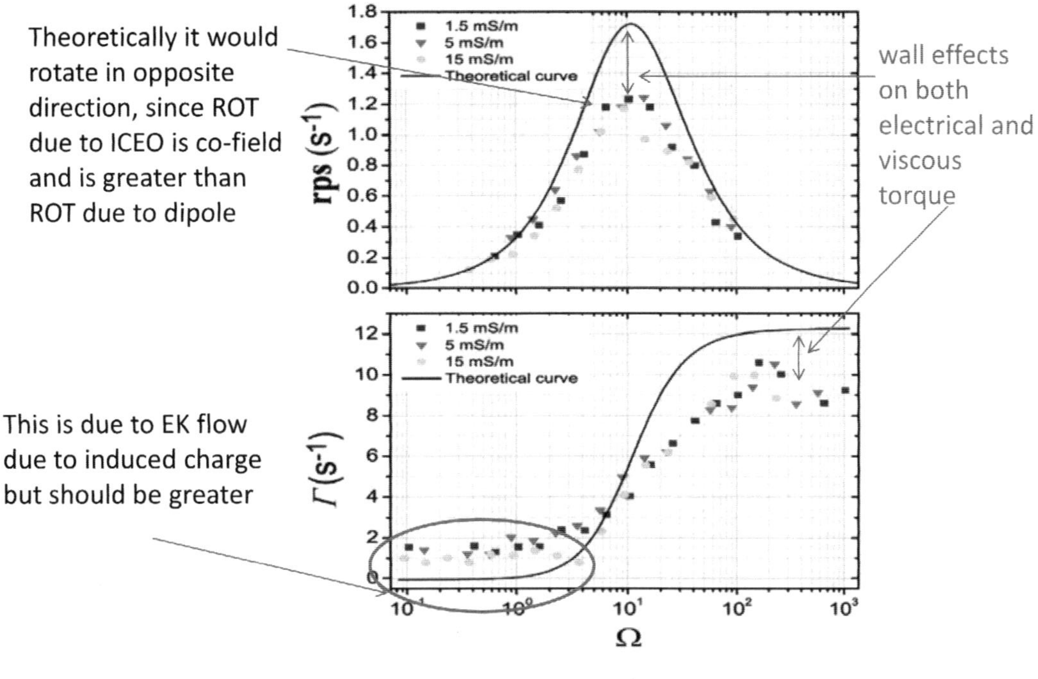

ICEO velocities observed experimentally are always smaller than standard theory for ideal EDL

(Ideal EDL is a double layer which is composed only of diffuse layer)

$$v_{\text{slip}} = \frac{\varepsilon}{\eta}\langle \Delta\Phi E_s \rangle$$

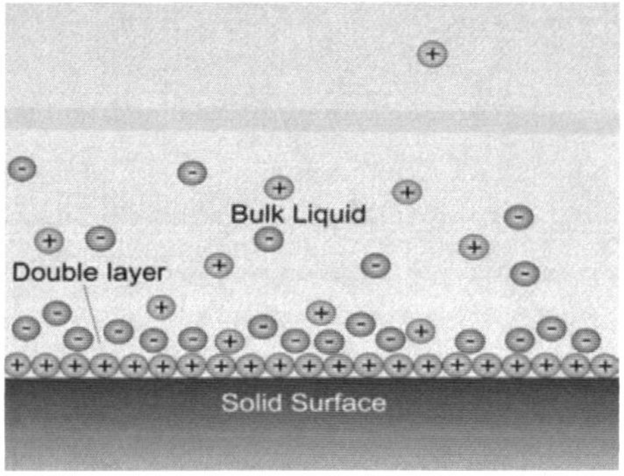

Theoretically, effects like ion **adsorption**, **oxide layers** on surface, surface **roughness**, **ion crowding**, or **surface conduction** reduce the Helmholtz-Smoluchowski velocity

$$\frac{E_0 a}{k_B T/e} \sim 1$$ => ion crowding may be disregarded

Outline

- Introduction: Metal particles in electrolyte
- DEP and ROT of microspheres
- EOr and ROT of nanowires
- Electrokinetic interactions between nanowires

Electrokinetic interactions between nanowires

Pair interaction at low frequencies

3.2 kHz
σ=1.5 mS/m

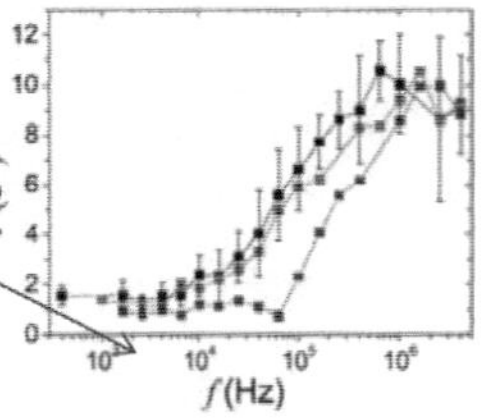

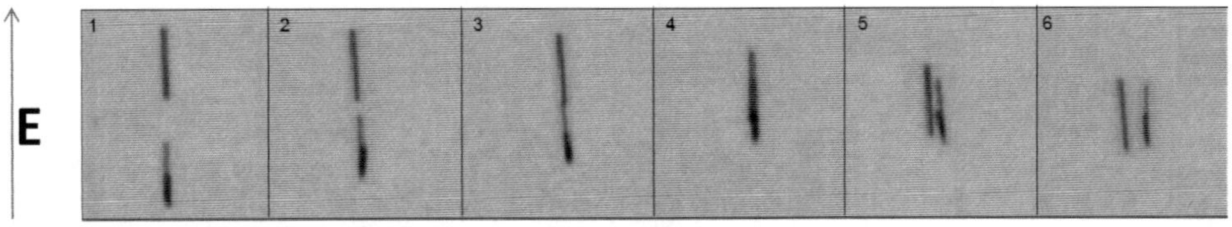

time

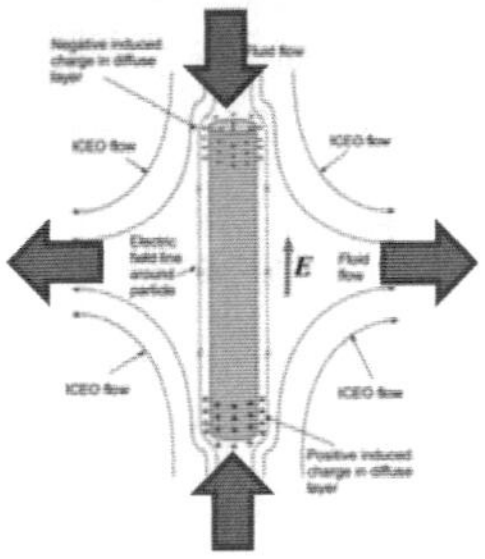

ICEO flow interaction

(Rose et al PRE 2009)
(Arcenegui et al PRE 20013)

Electrokinetic interactions between nanowires

Pair interaction at high frequencies

1.6 MHz
σ=1.5 mS/m

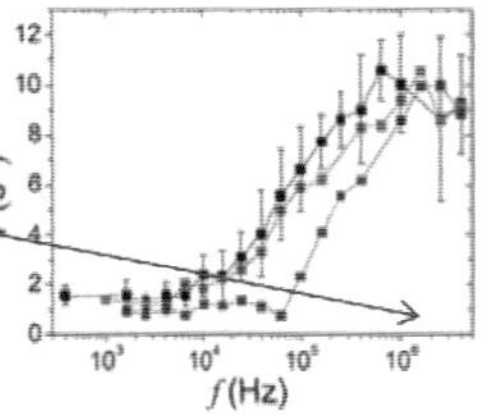

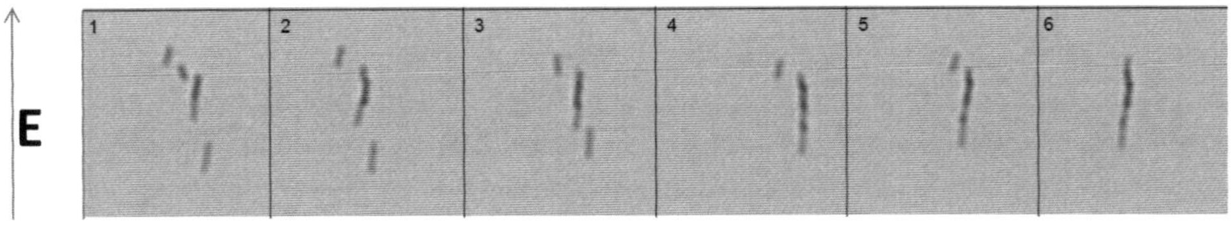

time

Pearl chaining
dipole-dipole interaction

Nanowires assembly induced by Electrokinetics as a function of frequency

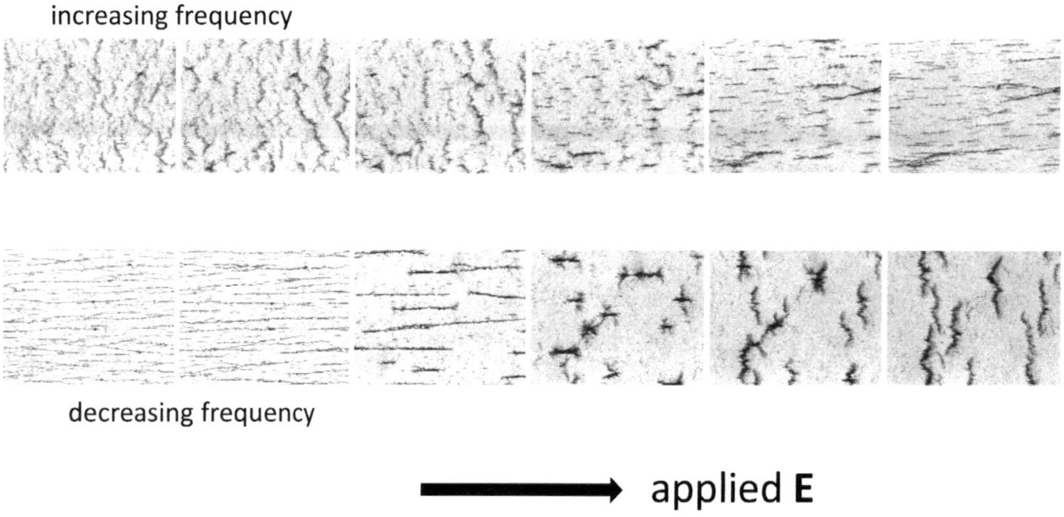

steady-state at different frequencies

Conclusions

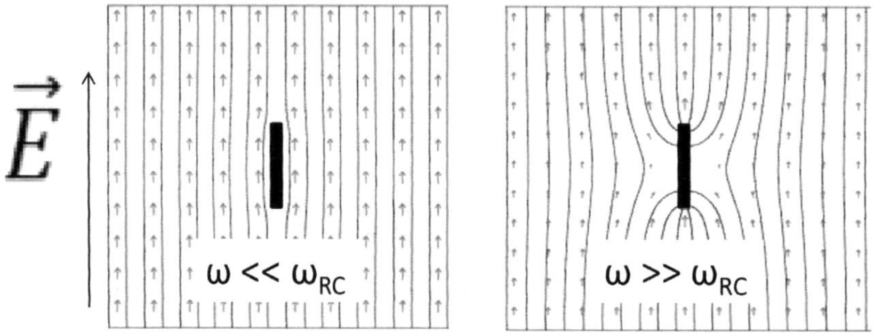

-Metal particles in solution behave as insulators at low frequencies and as conductors at high frequencies
- ICEO flow around particles affects particle motion at low frequencies, although it is smaller than predicted by standard theory of ideal DL
- Metal spheres and rods in solution are ideal to explore ac electrokinetics: comparison of low (ICEP) and high frequencies (dipole interaction)

438194 Targeted Proteomics with Single-Cell Resolution

Abstract:

Targeted Proteomics with Single-cell Resolution

Amy E. Herr, Department of Bioengineering, University of California, Berkeley

Innovation in measurement science is needed to transform pathology tools from subjective, qualitative visual inspection of biopsied tissues for just a handful of protein markers to objective, quantitative measurement of surface and intracellular signaling proteins, including (importantly) protein isoforms and sub-cellular location. In this talk, I will focus on my laboratory's recent efforts to measure multiple protein biomarkers to enhance tumor classification. Our overarching goal is to overcome limitations of workhorse immunohistochemistry (IHC) tools which rely on high-specificity antibody probes which are often not available. This presents an important analytical gap, as single-cell protein assays form the basis for cancer sub-type classification and treatment decisions.

Protein tools optimized for minute tissues and even single-cells are needed for three reasons. First, comparative analyses of intra-tumor regions (e.g., leading vs. interior regions <500 microns in diameter) elucidates resistance progression, which develops differently in different tumor regions. Conventional western blotting (WB) is not sensitive enough. Second, cell-to-cell variation is a hallmark of cancer; measurement of oncoprotein isoform heterogeneity is, in many cases, currently impossible. Third, as with all biobanks, limited tumor specimens are available. This is particularly true with HER2+ BCa, which is often biopsied early when tumors are small. Through microfluidic design strategies, we address both gaps by: single-cell resolution WB detection limits are 0.1% of conventional WB and well-matched to tissue region analysis. To achieve the lower limits of detection, we use an innovative strategy that differs from most microfluidic electrophoresis separations. We do not use a "cross-t" injector that injects <1% of the sample into the separation channel. Our distinct approach borrows from classic electrophoresis formats that pre-concentrate and then inject all material from the sample well into the separation lane. Minimizing dilution allows the scWB to detect protein from ~1 cell. Essentially we employ the same physics as flow cytometry which routinely and successfully measures single-cell protein levels. As with flow cytometry, the scWB limits dilution of cell contents; the protein concentrations in single mammalian cells can be quite high (i.e., nM to uM in ~1pL cell). Like flow cytometry, the scWB has a limit of detection of $\sim 10^4$ molecules.

Our single-cell WB platform is designed for both simplicity and performance. The assay performs $\sim 10^3$ concurrent single-cell Western blots in ~4 hours. The scWB device is a microscope slide coated with a thin layer (30 µm) of photoactive polyacrylamide gel. Using this simple device, the scWB assay steps include: (1) settling of single cells into microwells stippled in the gel layer (30 um diameter), (2) in-situ cell lysis, (3) polyacrylamide gel electrophoresis (PAGE), (4) photo-initiated blotting to immobilize proteins, and (5) antibody probing with scWB fluorescence readout using a microarray fluorescence scanner. PAGE is performed in ultra-short 500 um to 2 mm separation distances compared to 10's of centimeters for slab-gels.

As is directly relevant to next-generation pathology tools, we are applying the single cell targeted proteomic tool to assess oncoprotein isoforms. In particular, with our collaborators at the Stanford University School of Medicine, we measure truncated oncoprotein isoforms are not detectable via genomic or transcriptomic tools. The truncated proteins are created by proteolytic cleavage (ADAM10) and alternative translation, thus requiring direct protein measurement. While much attention has centered on nucleic acid assays in single cells and small tissue sections, paltry attention has focused on direct measurement of protein targets, a major innovation of our work. Immunoassays for truncated oncoprotein isoforms are impossible, as isoform-specific antibodies do not exist. While mass spectrometry is ideal for protein isoform detection, mass spectrometry is not sensitive enough for single-cell measurements. Consequently, immunoassays are the de facto standard for protein measurement in tissues down to single-cells. Yet, immunoassays fail if target-specific antibodies do not exist, as is the case for oncoproteins including HER2; available antibody reagents cannot distinguish the various truncated HER2 isoforms. Our isoform measurement innovation draws on microfluidic integration of multi-stage assays. We will measure proteins directly by employing separations first, and only after isoforms are resolved by size do we use antibody probes. Consequently, even moderate-to-poor performance antibodies are suitable, as separations mitigate poor antibody selectivity. While developed for HER2, tools for protein isoforms will be broadly relevant, as several driver oncoproteins are isoforms (e.g., RON33, MDM234, MYC35, BAG-136, PPM1D37, FLIP38).

Further, high background signal and off-target interactions limit the number of proteins detectable in the same, individual cell to ~5 (workhorse ICC & flow cytometry). Multiplexing to >10 proteins has been demonstrated, but employed uncommon tools that can be difficult to apply. To multiplex a Protein Panel, we use a dual-pronged approach: (1) create micro-gradient gels (similar to gradient gels) but created via gray-scale photolithography to achieve high quality microscale separations and (2) employ a novel photo-active polyacrylamide gel that supports protein electrophoresis (molecular sieving) then switches to a protein immobilization scaffold for subsequent immunoprobing. During brief UV illumination, benzophenone moieties in the polymer abstract hydrogen from C—H bonds on the protein, resulting in stable covalent protein immobilization. Covalent immobilization allows chemical stripping for tens of re-probing rounds thus allowing multiplexing up to 10's of protein targets

Classification of tumors suffers from tremendous pathologist-to-pathologist variability in interpreting IHC. Gene screens yield quantitative readouts, oncoprotein isoforms cannot be measured, and multiplexed signaling protein assays are not feasible. I will discuss these advances in microfluidic assay integration for high protein specificity even in small tissue samples and single cells as is directly relevant to precision medicine.

C-C Kang, J-M Lin, Z Xu, S Kumar & AE Herr. "Single-cell western blotting after whole-cell imaging to assess cancer chemotherapeutic response." Analytical Chemistry, 2014, 86(20):10429-36.

AJ Hughes*, DM Spelke*, Z Xu, C-C Kang, DV Schaffer, & AE Herr. "Single-cell western blotting." Nature Methods, 2014, 11(7):749-55.*Equal contributors

438234 Determining Reactive Oxygen and Nitrogen Species in Cells By Microchip Electrophoresis

Abstract:

Microchip electrophoresis is a powerful tool for the analysis of biological samples. In particular, the ability to perform fast efficient separations makes it possible to monitor multiple compounds simultaneously with high temporal resolution. The small sample volume requirements of the chip make it ideal for the analysis of microdialysis samples and single cells. Reactive oxygen and nitrogen species (RNOS), such as nitric oxide, peroxynitrite and superoxide, have been implicated in a number of cardiovascular and neurodegenerative diseases. RNOS are generated as part of the immune response and can react with DNA, proteins and lipids leading to cell death or dysfunction. Approaches for the detection and quantitation of multiple RNOS simultaneously are lacking, especially at the single cell level. In this presentation, methods for the detection of RNOS using microchip electrophoresis with electrochemical and laser induced fluorescence will be described. Many RNOS, including as nitrite, peroxynitrite, hydrogen peroxide and nitric oxide, are electroactive and can be separated and detected using microchip electrophoresis with electrochemical detection (MEEC). An advantage of MEEC is that intracellular oxidants, which play a protective role against oxidative stress (glutathione and ascorbate), can be measured under the same conditions. Alternatively, laser induced fluorescence detection (LIF) can be used for the indirect detection of RNOS including nitric oxide, superoxide and peroxynitrite. ME-LIF requires derivatization of the analytes with specific fluorophores prior to analysis. In this case, the electrophoretic separation is used to isolate the desired product from reaction side products, allowing for more precise quantitation. Examples of the detection of RNOS in macrophage cells using both ME-EC and ME-LIF will be presented.

Funding from the National Science Foundation is gratefully acknowledged.

438132 Non-Viral Gene and Biomolecule Delivery By Nanochannel Electroporation

Abstract:

Non-viral Gene and Biomolecule Delivery by Nanochannel Electroporation

L. James Lee[1,2,3,4,6], Daniel Gallego–Perez[2], Xi Zhao[1,2], Lingqian Chang[2,3], Keliang Gao[2], Veysi Malkoc[2], Nick Chiang[2,4], Xiaomeng Huang[2,4], John C. Byrd[4], Raj Muthusamy[4], Wu Lu[2,5] and Chandan Sen[6]

[1]Department of Chemical and Biomolecular Engineering

[2]Center for Affordable Nanoengineering of Polymeric Medical Devices

[3]Department of Biomedical Engineering

[4]Comprehensive Cancer Center

[5]Department of Electrical and Computer Engineering

[6]Center for Regenerative Medicine and Cell Based Therapy
The Ohio State University

The ability to deliver precise amounts of biomolecules and nanofabricated probes into living cells offers tremendous opportunities for biological studies and therapeutic applications. It may also play a key role in the non-viral generation of engineered stem cells and induce pluripotent stem cells with high efficiency and non-carcinogenic properties. Currently-available transfection approaches are heavily dependent upon diffusion- and endocytosis-based mechanisms, which results in highly stochastic transfection. Dosage controlled delivery to multiple cells is not achievable with any existing techniques. We have overcome this problem by developing a new technology, nanochannel electroporation* (NEP) allowing transfection of many small sized and delicate cells with precise control over dose and timing. Cell mortality from NEP is virtually zero. We show dose control effects on a variety of transfection agents such as oligonucleic acids for cancer therapy, molecular probes for intra-cellular biomarker detection and large plasmid for cell reprogramming.

We first describe the principle and design of a series of new NEP systems based on electrokinetic flows and electroporation forces. They include a 2D NEP-optical tweezers platform for single cell transfection*, a 3D DEP-NEP platform for large scale cell transfection, and a 3D NEP patch for *in vivo* gene transfection. For 3D NEP chips, a properly-engineered and massively parallel ordered array of Si nanochannels, capable of handling/transfecting ~60,000 cells/cm^2, was fabricated using cleanroom technologies. Positive dielectrophoresis (pDEP) was used to selectively place cells on the nanochannel outputs, thus allowing for controlled and highly efficient transfection. Single-cell dosage control capabilities were demonstrated using a wide variety of small and large molecules. We then present our on-going work on several important medical applications using those NEP systems.

Short interfering RNA (siRNA) has been widely used to silence its targets through complementary-binding to the target mRNA. The conventional delivery methods such as bulk electroporation and nanocarriers cannot provide uniform delivery and are stochastic in nature. The variation in delivery to each cell may cause different downstream consequences, making it difficult to understand the fundamentals of cell response to different drug dosages. An anti-apoptotic gene, MCL-1 was found elevated in some subset of Acute Myeloid Leukemia (AML) patients who harbor with FMS-like tyrosine kinase-3 (FLT3) mutation and have poor survival. We used our 2D NEP method to deliver MCL-1 siRNA to individual AML cells with controlled dosage, which revealed different thresholds of MCL-1 siRNA induced apoptosis to AML cells with or without FLT3 mutation. We were able to measure delivered MCL-1 siRNA and endogenous mRNA level and their dynamic changes by molecular beacon and single cell qRT-PCR and revealed the behavior of MCL-1 siRNA as a therapeutic agent at single cell level.

For the 3D DEP-NEP platform, the first example is the isolation and interrogation of living leukemic CLL stem cells *in vitro* from CLL patients. We used the 3D NEP technology to deliver molecular probes to identify rare cancer initiating cells with unique intracellular markers, and then interrogate their stemness and drug resistance *in vitro*, The second example is *in vitro* cell reprogramming. Although cell reprogramming holds great promise for regenerative medicine, safety concerns and/or the stochastic nature of viral or non-viral transduction approaches have hampered clinical translation. Here we report on the development of a novel non-viral nanotechnology-based platform that allows for deterministic large-scale cell transfection with single-cell resolution. The superior capabilities of such technology were demonstrated using a well-established induced neuron reprogramming model via overexpression of *Brn2, Ascl1, and Mytl1* (BAM). Our approach resulted in reprogramming performance similar to viral methodologies and much better than existing non-viral methods. Furthermore, increased neuronal complexity could be tightly tailored by varying the BAM ratio and substantially enhanced by adding patterning genes to the BAM cocktail because of the easy control of cocktail receipt and dosage. Although the mechanism remains obscure, we observed significant stroke recovery in mice when NEP induced neurons were implanted in the mouse brain. In addition, the high-throughput NEP system allows living cell interrogation of the reprogramming process. We discovered that BAM-mediated reprogramming is regulated by *Ascl1* dosage, is regulated by the S-phase cyclin *CCNA2*, and that at least some neurons induced through BAM-mediated neuronal reprogramming pass through a nestin-positive progenitor cell stage.

The NEP patch is designed for *in vivo* gene transfection. *In vivo* cell reprogramming has major clinical implications. Ideally, a patient's own tissue could be used as a prolific bioreactor to generate specific cells or tissue with minimal immunorepulsion and contamination. Today, non-viral reprogramming has rarely been achieved *in vivo* and little is known about the reprogramming mechanism within tissues. Herein we report a facile and yet powerful NEP patch technology to non-virally transfect and remodel/reprogram naturally-exposed or surgically-accessible tissue surfaces. Experiments with mouse skin and muscle show that by transfecting combinations of reprogramming genes into the outermost cell layers, secondary mechanisms could be triggered to propagate such transfection throughout the tissue via paracrine-like signaling, presumably mediated by extracellular vesicles (EVs) pre-loaded with a plethora of bioactive factors in addition to the transfected gene mRNAs. Overexpression of these genes led to tailored tissue remodeling and localized cellular reprogramming into different fates, e.g. induced neurons, brown adipocytes, and endothelium, depending on the specific gene combination used. Finally we demonstrate that skin transfection-activated reprogramming (STAR) could effectively revascularize ischemic tissues and limbs in mice.

*P. E. Boukany, A. Morss, W-C Liao, B. Henslee, X. Zhang, B. Yu, X. Wang, Y. Wu, H.C. Jung, L. Li, K. Gao, X. Hu, X. Zhao, O. Hemminger, W. Lu, G. Lafyatis and **L.J. Lee**, "Nanochannel Electroporation Delivers Precise Amounts of Biomolecules into Living Cells", Nature Nanotechnology, 6, 747-754 (2011), research highlight in Nature Methods, 8, 996-997 (2011).

438810 A Conic Nanopore Array for Point-of-Care Microrna Profiling

Abstract:

We report a low-cost and turn-key Point-of-Care (POC) membrane-nanopore hybrid platform that can identify and enumerate a panel of 20-nucleotide microRNAs (miRNA) in a heterogeneous sample, with minimum pretreatment and with a dynamic range comparable to rtPCR. The approach is to integrate in a multi-stage platform existing membrane microfluidic pretreatment technologies with new solid-state conic nanopore array technologies. New nanofluidic technologies have been used to offer ultra-fine molecular isolation and enumeration capabilities: nanoparticle assembly for probe loading and non-target gating, pore surface modification to prolong translocation of short miRNA for intrapore hybridization capture and pore level ion-current Ohmic heating for precision melting of hybridization complexes to release a small number of targets. The throughput is one-thousand to one-million times higher than a single nanopore and yet the translocation time of individual miRNA is sufficiently long for hybridization. This integrated multi-stage design allows us to isolate and interrogate a minority subset of short miRNAs from a large population of molecules in a heterogeneous sample. It complements our membrane sensor design to offer a limit of detection of 100 copies and 7 decades of dynamic range.

413050 Single-Molecule Translocation through Asymmetric Conic Nanopores: The Effects of Van Der Waal Absorption and Electro-Osmotic Flow

Abstract:

We examine the pertinent electrokinetics of single-molecule translocation through conic nanopores both experimentally and theoretically. Due to the ion-selectivity of the nanopore and the conductivity gradients introduced by the spatial asymmetry, anomalous ion current phenomena such as rectification, transient hysteresis and water-splitting have been observed. In our past work, we have derived asymptotic expressions for the rectification factor, the hysteresis window and the water-breaking cyclic voltammetry signatures with a perturbation analysis of the area-averaged ion transport equation. We use optimum designs from such theories to deliver short (20 bases) DNAs to a conic nanopore and to delay their translocation time, two major obstacles for nanopore rapid sequencing technologies. Using non-equilibrium concentration and charge polarization endowed by the asymmetry, we are able to produce an external electro-osmotic flow that linearizes, convects and focuses the DNA molecules into the nanopore. At the pore tip, van der Waal forces between the pore and the translocating molecule are shown to be stronger than Coulombic repulsion after surface treatment with a high-permittivity material. Curiously, short 20b single-stranded DNAs show more absorption affinity than their double-stranded counterparts of the same length. This is attributed to the exposed ring structures of ssDNA with delocalized electrons that enhance the London dispersion attraction to the tip surface. We present Poisson distributions of both ssDNA and dsDNA with average translocation times that are separated by two orders of magnitude. We also show ion-current data and super-microscopy images that suggest a stick-slip dynamics for the translocating ssDNA.

Yan, Y., Wang, L., Xue, J., and Chang, H.-C., "Ion Current Rectification in ConicNanopores: Nonequilibrium Ion Transport Biased by Ion Selectivity and Spatial Asymmetry", J Chem. Phys., 138, 044706 (2013).

Liu, S., Yan, Y., Wang, Y., Senapati, S. and Chang, H.-C., "Plasmonic Hotspots of Dynamically Assembled Nanoparticles in Nanocapillaries: Towards a miRNA Profiling Platform", Biomicrofluidics, 7, 061102 (2013).

Conroy, D. T., Craster, R. V., Matar, O. K., Cheng, L.-J. and Chang, H.-C., "Non-Equilibrium Hysteresis and Wien Effect Water Dissociation at a Bipolar Membrane", Phys Rev E, 86, 056104 (2012).

Adapting current monitoring measurement for low conductivity media

Mario A. Saucedo-Espinosa and Blanca H. Lapizco-Encinas*

+Microscale Separations Laboratory, Rochester Institute of Technology, NY, United States

Introduction

Microfluidics studies the design and application of systems where small volumes of liquid are handled. Many of the materials commonly employed in microfluidics acquire electric surface charges upon contact with an electrolyte solution and develop an electrical double layer (EDL) [1]. The interaction of the EDL with an externally applied DC electric field results in an electroosmotic (EO) flow [1], which is frequently employed to pump fluid and particles through microfluidic channels. Electroosmotic flow is attractive since it produces a plug-like velocity profile, causing low sample dispersion and ease of application.

A common method to estimate the average EO velocity in a microchannel is current monitoring [2], which offers the advantages of being simple and cost-effective to implement. The technique makes use of the electric current changes when a test electrolyte solution is replaced by an auxiliary solution, whose concentration is different. Current monitoring has been applied to estimate the EO mobility in glass [3] and polymer [4] microchannels. Usually, the method employs a straight channel with a known length, and the time that the test solution takes to be completely displaced is measured. The average EO velocity is calculated simply by dividing the channel length by the measured time. However, this approach requires a constant interaction of the user with the experimental setup [5], which can introduce noise and error in the estimation of the EO velocity. As an example, the test solution must be removed completely at the inlet reservoir and replaced by the auxiliary solution. If the test solution is not completely removed, significant mixing can arise, which affects the measurement of time required for the displacement in the channel to occur [5]. Therefore, current monitoring measurements are

* Email: bhlbme@rit.edu

commonly repeated several times before an accurate assessment can be obtained [5]. To overcome this disadvantage, Sze *et al.* [6] proposed to use the slope of the current-time relationship during the channel displacement to estimate the average EO velocity. This is a more robust method, since it is less affected by the noise at the beginning and ending sections of the displacement process. The authors reported zeta potentials for glass and PDMS which were reproducible within ±6%.

Practical difficulties arise, however, when the ionic concentration of the test solution is small. This is the case of low conductivity suspending mediums used in insulator-based dielectrophoresis (iDEP) [7]. Given that a small concentration difference between the test and auxiliary solutions is needed, the signal-to-noise ratio of the signal to measure decreases. Hence, measuring the current change with sufficient accuracy becomes challenging.

This work presents the advantages of using a system with parallel channels to increase the signal-to-noise ratio of the current signal to measure. Further, the proposed method allows the use of all the experimental data generated on multiple independent experiments to estimate a single EO velocity; keeping all of the current monitoring advantages and extending its use for low conductivity media.

Theoretical background

Electroosmotic flow

Electroosmosis is the motion of an electrolyte solution when a DC electric field is applied in a channel having the capability of developing an EDL. The ions within the EDL acquire momentum and exert a drag force to the fluid. The Helmholtz-Smoluchowski equation describes the EO velocity in the limit of a thin double layer [1]:

$$\vec{v}_{eo} = \mu_{eo}\vec{E} = -\frac{\varepsilon_m \zeta_w}{\eta}\vec{E}, \tag{1}$$

where μ_{eo} is the EO mobility, defined in terms of the real permittivity and viscosity of the suspending medium, ε_m and η, respectively, and ζ_w is the zeta potential of the channel wall.

Current-time analysis

In the cases where the ionic concentration difference between the test and auxiliary solutions is small, the time response of the current is linear [2]. The slope based method, modified to account for the conductivity of the channel walls, is [2]:

$$\mu_{eo} = \frac{AL}{E_x \cdot (I_T - I_D)}, \tag{2}$$

where A is the slope of the current-time plot (Fig. 1), L is the length of the displacement channel and E_x is the applied electric field. Further, I_T and I_D are the electric current plateaus when the displacement and main channels are filled with the test and auxiliary solutions, respectively (see Fig. 1).

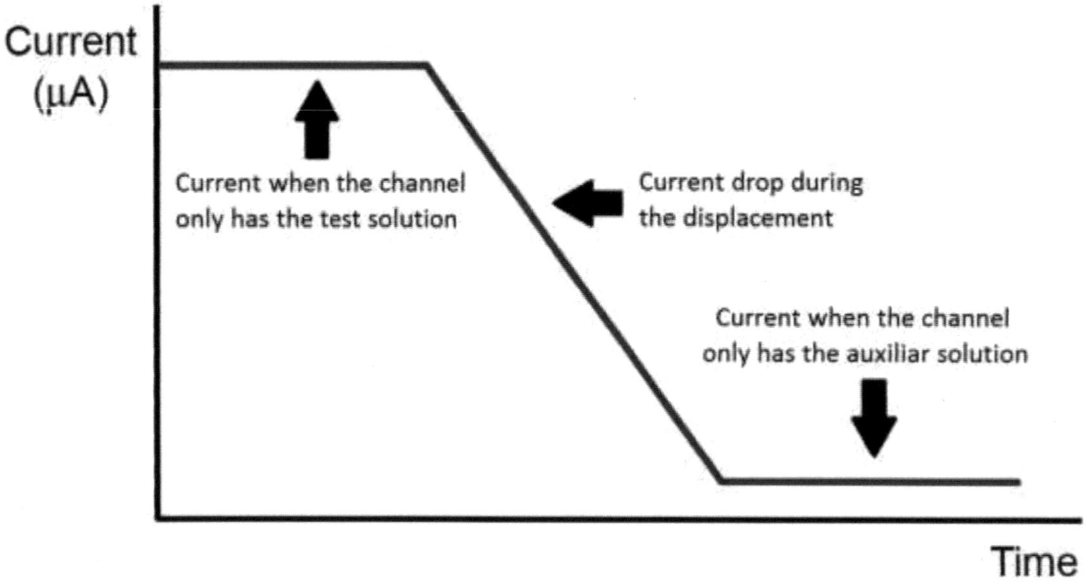

Figure 1. Illustration (noise-free) of the time response in a current monitoring setup. Initially, the channel is full of the test solution, and the current is stable. Next, the test solution at the inlet is replaced for an auxiliary solution and the displacement process starts, causing a linear decrease in the current. Finally, the channel is full of the auxiliary solution and the current reaches a second steady state value.

The zeta potential of the channel walls, ζ_{wall}, can be calculated using the EO mobility:

$$\mu_{eo} E_x = -\frac{\varepsilon_m \zeta_{wall}}{\eta}, \qquad (3)$$

where ε_m and η are the permittivity and viscosity of the test solution.

Experimental section

Equipment

A high voltage sequencer (Model HVS6000D; LabSmith, Livermore, CA) was used to apply DC electric potentials by means of platinum wire electrodes. The voltage sequencer was manipulated with the software Sequence provided by the manufacturer, which also incorporates a module for monitoring the current in the system.

Microdevices

Microchannels were made from PDMS (Dow Corning, Midland, MI) and fabricated using standard soft-lithography techniques. Each microchannel had two inlets and an outlet liquid reservoir. The displacement channel of each device was 5-cm long, 1-mm wide and 30-μm deep. The measurements were performed in systems with three channels working in parallel.

Results and discussion

The time response of the current is comprised by three main behaviors: (1) an initial stable signal when the channel is full of the test solution, (2) a linear decay in the current during the displacement process and (3) a second stable signal when the channel is full of the test solution. Figure 2 presents the time response of the current for a single channel. The test solution (the initial solution in the channel) is 0.1 mM KCl (pH = 6.3 and σ = 16.3 μS/cm). The auxiliary solution (the one that displaces the test solution) was prepared by diluting 9 parts of the test solution with one part of DI water, for a final ionic concentration of 90% the test solution. As observed, the noise in the measured signal hinders the stable states of the current, and poses a challenge in the determination of when the displacement process starts and finishes. This uncertainty complicates the estimation of the slope needed in Eq. (2) when estimating the EO mobility. For instance, the estimation of ζ_{wall} for the single channel system with Eq. (3) results in ζ_{wall} = -98.7 ± 8.5 mV. In comparison, the ζ_{wall} under similar experimental conditions was determined to be -110 ± 6.6 mV [6]. However, several experiments were needed before the successful signal presented in Figure 2. The R^2 coefficient for the linear regression employed to estimate the slope is R^2 = 0.28. This indicates that the signal-to-noise ratio is small, hence, the noise hinders the real value of the slope and several experiments are needed until the slope can be adequately evaluated.

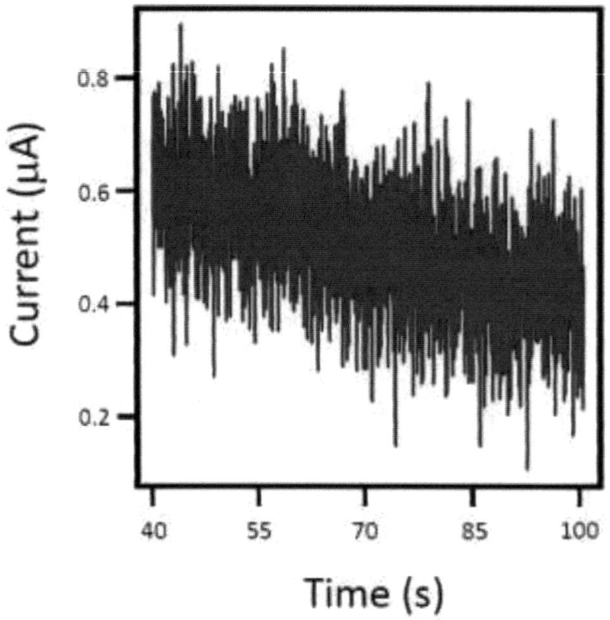

Figure 2. Linear decrease in the current during the displacement process in the single channel system. The signal-to-noise ratio is relatively small.

On the other hand, Figure 3 shows the time response of the current when three channels are used in parallel. Given the parallelization, all channels are subjected to the same electric field, and the current is measured in the three channels as a whole. This increases the measurable signal, which increases the signal-to-noise ratio. As observed in Figure 3, the three regimes of the current are more distinguishable when compared to using a single channel. The two stable regions can be discerned from the points composing the linear decay in the current. The estimation of ζ_{wall} for the parallel system results in ζ_{wall} = -105.2 ± 1.2 mV. The confidence interval of ζ_{wall} is narrower when using the parallel system. In addition, less experiments were needed to arrive to this result, and the linear regression coefficient was calculated as R^2 = 0.94. This value indicates that the signal-to-noise ratio increased in the parallel system, ultimately improving the estimation of the slope in the time response of the current.

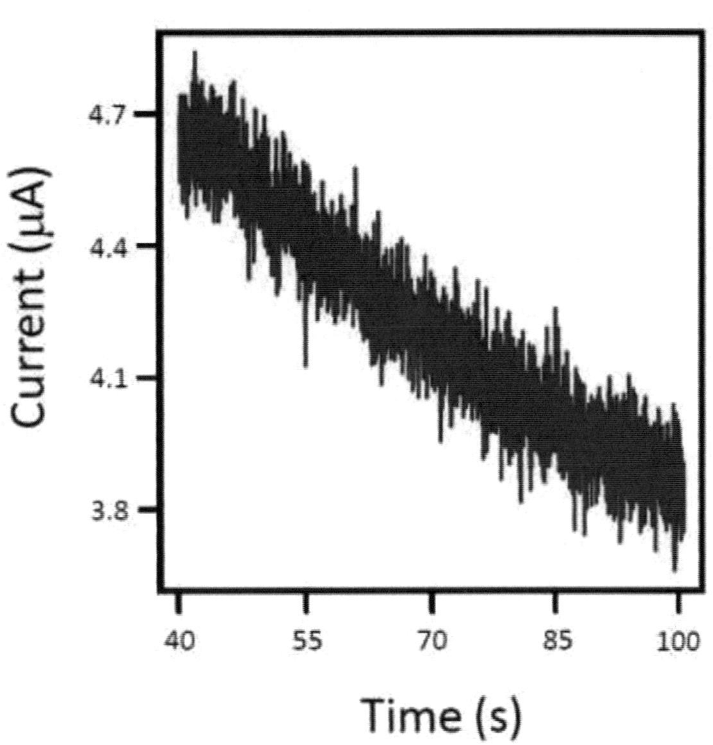

Figure 3. Linear decrease in the current during the displacement process in the parallel channel system. The signal-to-noise ratio is relatively large.

Conclusion

This study presents an enhancement of the current-time analysis for the application of the current monitoring technique in low conductivity systems by using parallel channels. The parallelization presented here allows the use of all the experimental data to estimate an electroosmotic velocity, increasing the signal-to-noise ratio. The results show the potential that this technique has to effectively estimate the EO velocity in low conductivity media.

References

[1] Probstein, R. F., Physicochemical hydrodynamics: an introduction, John Wiley & Sons, 2005.

[2] Almutairi, Z., Ren, C. L., Simon, L., *Colloids and Surfaces a-Physicochemical and Engineering Aspects* 2012, *415,* 406-412.

[3] Huang, X., Gordon, M. J., Zare, R. N., *Analytical Chemistry* 1988, *60,* 1837-1838.

[4] Locascio, L. E., Perso, C. E., Lee, C. S., *Journal of chromatography A* 1999, *857,* 275-284.

[5] Almutairi, Z. A., Glawdel, T., Ren, C. L., Johnson, D. A., *Microfluidics and Nanofluidics* 2009, *6,* 241-251.

[6] Sze, A., Erickson, D., Ren, L., Li, D., *Journal of Colloid and Interface Science* 2003, *261,* 402-410.

[7] Aldaeus, F., Lin, Y., Roeraade, J., Amberg, G., *Electrophoresis* 2005, *26,* 4252-4259.

[8] Björck, A., Numerical methods for least squares problems, Siam, 1996.

428752 Field Effect Control of Ion Transport in a pH-Regulated Nanochannel with Overlapped Double Layers

Abstract:

The ability to control the ion transport, which depends exquisitely on surface charge properties and ionic concentration distributions, in nanofluidic devices plays a dominant role for emerging applications such as energy conversion, ionic diodes, and sensing of molecules. To achieve active control, nanofluidic field effect transistors (FETs), referring to gate electrode-embedded nanofluidic devices, have been developed recently. To improve the development of FET control nanofluidic devices for emerging applications, analytical expressions taking into account practical effects of the presence of hydrogen and hydroxide ions, electroosmotic flow, Stern layer effect, and surface chemistry reactions on the channel wall, are derived to study the field effect control of surface charge property and ionic conductance in a pH-regulated nanochannel with overlapped electric double layers (EDLs). We show that the EDL overlap effect is relatively significant at low pH and salt concentration when a negative gate potential is applied. If pH is low, the EDL overlap effect on the field control of zeta potential of the nanochannel is remarkable at large positive gate voltage, while that effect on ionic conductance is significant at large negative gate voltage.

429319 Taylor Dispersion in Electrofocusing at Steady-State

Abstract:

The Aris-Taylor formulation for the dispersion coefficient should not be applied to electrophoretic separations where solutes are focused to a steady-state. Analytic expressions and numerical calculations for dispersion coefficients will be presented for several types of separations where a steady-state or pseudosteady-state exist. In the simplest case where a solute is focused against a linear gradient in a restoring force, dispersion is controlled by two dimensionless groups, a Peclet number, Pe, which characterizes the magnitude of the parabolic component of the flow, and a gradient number, G_f, which depends on the magnitude of the slope of the restoring force. Following a brief summary of the analytic expressions for dispersion in mathematically linear separations, this talk will discuss the nature of dispersion in nonlinear electrophoretic systems.

430468 Measurements of the Transverse Migration of Polyelectrolytes in Microfluidic Channels Induced By Combined Electric and Flow Fields

Abstract:

In a straight microfluidic channel, a dilute polyelectrolyte solution such as DNA can be focused towards the center of the device by applying an electric field and pressure gradient simultaneously. We report on systematic measurements of the effects of flow velocity, electric field, and ionic strength on the degree of migration. Our experiments show that migration increases with increasing shear and electric field, but eventually saturates; the addition of salt reduces the strength of the migration. Comparisons are made between experimentally measured concentration profiles and numerical models based upon the hypothesis that migration is caused by an electrically induced hydrodynamic interaction.

431314 Moderately Nonlinear Diffuse Charge Dynamics Under an AC Voltage

Abstract:

The response of a symmetric binary electrolyte between two parallel, blocking electrodes to a moderate amplitude AC voltage is quantified. The diffuse charge dynamics are modeled via the Poisson-Nernst-Planck (PNP) equations for dilute solutions of point-like ions. The solution to these equations is expressed as a Fourier series with a voltage perturbation expansion for arbitrary diffuse layer thickness and AC frequency. Here, the perturbation expansion in voltage is in powers of V_o/V_T, where V_T is the thermal voltage and Vo is the amplitude of the driving voltage. We show that the response remains essentially linear in voltage amplitude at frequencies greater than the *RC* frequency of diffuse layer charging ($D/\lambda L$, where D is the ion diffussivity, λ is the diffuse layer thickness, and L is half the cell width) but becomes nonlinearly voltage dependent at frequencies below the *RC* frequency. We find that the ion densities exhibit symmetric deviations from the equilibrium density at even orders of the voltage amplitude. This leads to the voltage dependence of the current in the external circuit arising from the odd orders of voltage. We use this to derive a nonlinear generalized impedance for moderate voltages. The first correction is a second order voltage dependent term which predicts a decrease in the complex part of the impedance. This is consistent with the predicted increase in the static diffuse layer capacitance from Gouy-Chapman theory.

433056 Resolving Overlimiting Current Mechanisms in Micro-Nanochannel Interface Devices

Abstract:

We present results demonstrating the space charge-mediated transition between classical, diffusion-limited current and surface-conduction dominant over-limiting currents in a shallow micro-nanochannel device [1]. The extended space charge layer develops at the depleted micro-nanochannel entrance at high current and is correlated with a distinctive maximum in the dc resistance. Experimental results for a shallow surface-conduction dominated system are compared with theoretical models, allowing estimates of the effective surface charge at high voltage to be obtained. Further, we extend the study to microchannels of moderate to large depths where the role of various electro-convection mechanisms becomes dominant. In particular, electro-osmotic of the second kind and electro-osmotic instability (EOI) which competes each other at geometrically heterogeneous (e.g. undulated nanoslot interface, array of nanoslots) nanoslot devices [2]. Also, these effects are shown to be strongly modulated by the non-ideal permselectivity of the nanochannel.

References

[1] J. Schiffbauer, N. Liebowitz, U. Liel, S. Park, and G. Yossifon, Probing space charge and resolving overlimiting current mechanisms at the micro-nanochannel interface, arXiv:1409.4548.

[2] Y. Green, S. Park, G. Yossifon, Concentration Polarization and Electroconvection in a Nanoslot Array of Varying Interchannel Spacing, Phys. Rev E. 91, 011002(R) (2015).

(565a) Trapping and Sorting of Bacteria in an Optoelectric Trap

Mishra, A.,
Maltais, T.,
Walter, T. M.,
Wei, A.,
Kinzer-Ursem, T. L., Purdue University
Williams, S. J., University of Louisville
Wereley, S. T., Purdue University

Originally presented on: 11/11/2015 13:00:00 - 13:15:00

Since the advent of microfluidics, non-contact particle trapping methods have been highly desired. Rapid Electrokinetic Patterning (REP) method, developed at Purdue University, has received considerable attention for controlled trapping and translation of particles. REP utilizes laser induced toroidal electrothermal flow and particle-electrode interaction for trapping particles. So far, REP has been predominantly used for trapping of abiotic micro- and nanoparticles. In this study, we demonstrate trapping of peritrichously flagellated, motile *Enterobacter aerogenes* cells. Conventionally, optical tweezers are used for trapping of motile bacteria. Due to motility of cells optical tweezers utilize intense laser beams which can induce photodamage to the bacteria. We experimentally demonstrate that REP can trap motile cells with four times lesser intense beam. In REP, trapping of bacteria is dependent upon their physical and dielectric properties. It allows us to sort tube shaped *Escherichia coli* cells from spherical *Staphylococcus aureus* cells.

421980 Dielectrophoresis in Porous Media: Electrically Switchable Filtration

Abstract:

Dielectrophoresis in Porous Media: Electrically Switchable Filtration

Georg R. Pesch, Fei Du, Jorg Thöming, Michael Baune

Chemical Engineering -- Recovery and Recycling (VdW), Department of Production Engineering and Center for Environmental Research and Sustainable Technology, University of Bremen

Dielectrophoresis (DEP) describes the movement of charged or uncharged matter in inhomogeneous electric fields and is a versatile and promising technique for particle manipulation. It is caused by the accumulation of polarization charges at material interfaces between the particles and the surrounding medium. The DEP force acting on particles is directly related to the electric field strength and its spatial change and depends mainly on the particle volume. Hence, the movement of very small particles (i.e., nanoparticles) requires vast electric field gradients. These gradients could be achieved by either using technically mature electrode configurations or by letting an originally homogeneous electric field scatter at solid obstacles, as it is the case in insulator-based (electrodeless) DEP (iDEP).

In the presence of an excitatory field, an obstacle of different material than the liquid medium will polarize. This effect is caused by charge separation induced by the excitatory field in combination with the different dielectric properties of the liquid medium and the obstacle's material. The resultant polarization charges itself will cause a highly inhomogeneous electric field, which is very suitable for targeted particle motion.

We applied this technique in the concept of dielectrophoretically driven filtration (cf. Fig 1) [1]: In this, a conventional deep-bed filtration process is altered so that the filter material is embedded between two electrodes. Because the pores of the filter are much larger than the target particle diameter, no noteworthy mechanical retention of particles occurs in the absence of an electric field. When applying an ac voltage on the electrodes, however, the generated electric field scatters at the porous dielectric. This results in a highly inhomogeneous field inside of the pores with maximum electric field values at their surface. These field inhomogeneities give rise to a DEP force acting on particles passing through the pores. It will direct them towards the pore surface, where the particles are trapped and reside until the electric field is turned off. Hence, we can separate particles in the nm-size range without filter cake formation by using filters with pore sizes being several orders of magnitude larger than the particle diameter. This effectively reduces pressure loss in comparison to conventional filtration processes and gives the possibility to easily turn off the filtration effect by switching off the electric field, enabling safe and easy resuspension of trapped NPs. In semi-continuous operation mode we reached separation and recovery efficiencies for dp = 340 nm particles in dpore 100 µm pores of up to 65% with a flow rate of 90 L h-1 m-2, as a proof of principle.

In this first experiment, module design and choice of filter material was not based on deep insight into dielectrophoretic effects in porous structures. In order to better understand the trapping and scattering mechanisms inside the pores we abstracted the originally complex 3 dimensional porous geometry to a quasi 2 dimensional flow channel containing an array of post-like field obstacles; a geometry which very much resembles insulator-based DEP flow channels. We developed a novel method to analyze and evaluate the polarization of such obstacles. It is based on the theory that the polarization of quasi 2-d obstacles in homogeneous electric fields can be expressed by superposition of an infinite amount of multipoles. The multipole coefficients can be extracted by any solution for the electrostatic potential by numerical integration. This method allows easy reproduction of the polarization field after the coefficients have been extracted. The possibility to compare the coefficients of different structures yields an easy way to benchmark the obstacles for DEP particle trapping.

For the first time, a thorough analysis on the influence of the obstacles cross-sectional geometry and material on their DEP particle trapping efficiency is conducted. From the polarization field obtained by the multipole coefficients we derive an equation for the DEP force exerted on spherical particles (based on the point-dipole approximation). With this equation and a superimposed fluid flow we are able to effectively describe effectiveness of a given obstacle for attracting and trapping particles out of a fluid stream using positive DEP (cf. Figs. 2 and 3). With this method we are able to derive crucial parameters for the dielectrophoretical filtration process, such as ideal flow rate, pore radius, porosity, pore geometry, and electric field strength for a given particle diameter. Additionally, this enables us to predict the separation efficiency and transfer function of a given filter-particle--system.

[1] G. R. Pesch, F. Du, U. Schwientek, C. Gehrmeyer, A. Maurer, J. Thöming, M. Baune: Recovery of submicron particles using high-throughput dielectrophoretically switchable filtration. Sep. Purif. Technol. 132 (2014) 728--735.

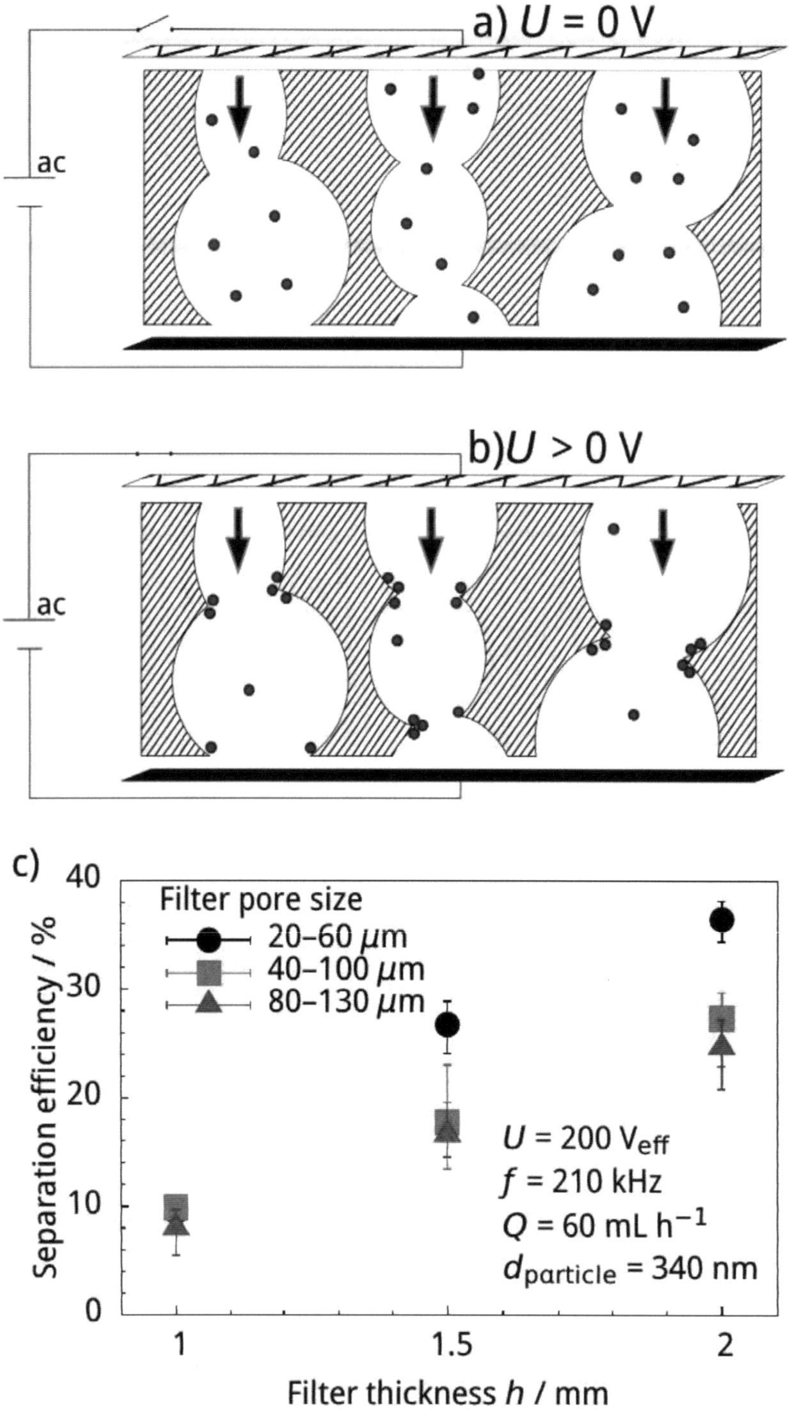

Figure 1: a) and b) show a sketch of the DEP filtration system. a) No voltage is applied and the particles simply flow through the comparably large pores. b) When sufficient voltage is applied, a DEP force is acting on the particles and directs them towards the surface oft he pores. C) Particle separation efficiency as a function of filter pore size and filter thickness at a specific flowrate of qÂ =Â 40 LÂ h-1 m-2.

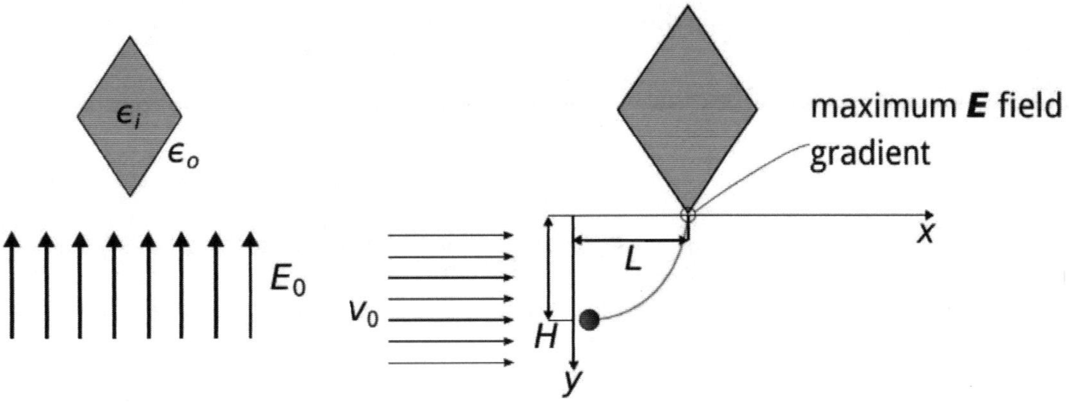

Figure 2: Post-like obstacle (as seen from top). Electric field applied perpendicular to the longitudinal axis. Particles are passing by the obstacle in distances H and L and are carried by a stationary fluid with velocity v0.

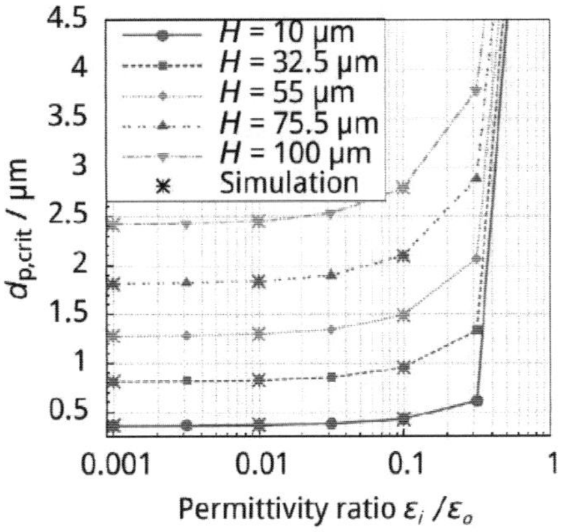

Figure 3: Critical particle diameter dp,crit which could be separated from a stationary fluid flow by a polarized obstacle with circular cross-section for different particle-to-obstacle distances H and different ratios between the obstacle's permittivity and the surrounding liquid medium's permittivity (cf. Fig 2).

429054 Comprehensive Micro RNA Profiling By Nanoparticle Electrophoresis in a Gel-Membrane Medium

Abstract:

Irregular expression of about 100 regulatory microRNAs (22 nucleotides) has been identified as viable disease biomarkers. Each disease requires quantification of a subset of these miRNA, with overlapping subsets for different diseases. With current rtPCR techniques, miRNA profiling for single or multiple diseases needs to be done with multiple samples whose cross-correlations are established with a common reference miRNA that is not irregularly expressed. Such multi-sample cross-correlation assays require exhaustive analysis with large compounded errors and have hence prevented miRNA profiling as a viable comprehensive screening test for multiple diseases. We report a novel nanoparticle gel electrophoresis technique that can potentially allow rapid quantification of 100 miRNAs from a single sample. A microfluidic platform for hybrid nanoparticle-gel electrophoresis isolates the miRNA population from the genome of a single lysed cell sample and interrogates them for multiple miRNA targets with minimum analyte loss. A high-field gel electrophoresis technique segregates short miRNAs from longer nucleic acids extracted on-chip from lysed cells or exosomes [1]. The isolated miRNA are then mixed on-chip with a family of nanoparticles of different size and fluorescent/Raman/plasmonic labels followed by preconcentration using an ion-exchange membrane. Dimer particles linked by the target have a lower electrophoretic mobility than the unlinked single particle, thus creating a new band whose intensity allows us to quantify the number of targets. Different particle labels and sizes then allows quantification of a large library of miRNA from a high-density nanoparticle ladder enabled by several new nanoparticle gel electrophoresis techniques.

[1] Taller, D., Richards K., Slouka, S., Senapati, S., Hill, R., Go, D. B. and Chang, H.-C., "On-Chip Surface Acoustic Wave Lysis and ion-exchange membrane detection of exosomal RNA for pancreatic cancer study and diagnosis", LabChip , 15, 1656-1666 (2015).

427717 The Effect of the Structure of Ion-Exchange Membranes on Ion Concentration Polarization and Associated Electrokinetics

Abstract:

The phenomena resulting from ion concentration polarization occurring in ion selective systems under DC voltage are increasingly used in the field of microfluidics to develop novel diagnostic and screening platforms with the main applications in point-of-care testing. Ion exchange membranes that are traditionally employed in electromembrane separation processes as electrodialysis and electrodeionization are one example of a system exhibiting ion concentration polarization. They have recently been used for pretreatment of biological samples, preconcentration of biomolecules and their specific detection [1, 2]. The ion exchange membranes are commercially available and their chemical and physical composition is tailored for the main application in electromembrane separation. The composition of the membranes, however, affects the exhibited behavior when in DC electrical field and it can play an important role in designing point-of-care microfluidic platforms. Our work attempts to find important links between ion exchange membrane structure and the exhibited behavior under various experimental conditions. We use Micro X-ray computed tomography to analyze small pieces of commercially available ion exchange membranes in a state when they are ready for electrochemical characterization (fully swollen membranes in a given electrolyte). The 3D analysis of the membrane focuses on the composition of the membrane surface that is in contact with an electrolyte since the interaction of the membrane surface and the adjacent electrolytes especially on the depletion side of the membrane controls the behavior of the whole system. We further characterize these small pieces of the membrane by performing standard electrochemical measurements and by running fluorescent imaging of the membrane and adjacent layers of electrolyte to capture the ion concentration polarization and resulting electrokinetic phenomena. The electrokinetic phenomena that are responsible for the occurrence of an overlimiting current are still a subject of scientific discussions and depend on the type of the membrane and the experimental set up. Our goal is to describe the effects of the membrane surface on electrokinetics of ion exchange membranes and use the knowledge to design optimal structure of ion exchange membrane for a given function in point-of-care systems.

[1] Senapati, S., Slouka, Z., Shah, S. et al. An ion-exchange nanomembrane sensor for detection of nucleic acids using a surface charge inversion phenomenon, *Biosensor&Bioelectronics,* 2014, 60:92-100

[2] Slouka, Z., Senapati, S., Chang, H.C., Microfluidic systems with ion-selective membranes, *Annual Review Analytical Chemistry*, 2014; 7:317-335

433803 Microfluidic Free Flow Electrophoresis Using Conductive PDMS Polymer Membranes

Abstract:

Free flow electrophoresis (FFE) utilizes a combination of fluid flow and electric field-induced electrophoresis to deflect analyte in directions perpendicular to the flow stream for sample separation or concentration. Since the first miniaturized FFE device was reported in the early 1990s, several microfluidic-based FFE systems have been developed for biomedical, analytical and diagnostic applications. One major issue with miniaturized FFE systems is that the electric field-generating electrodes must be isolated from the main flow stream to prevent on-chip bubble generation and electrode damage from water electrolysis. Many strategies are used to isolate the electrodes from the main flow channel, including conductive gels and side channel electrode arrays. These techniques, however, suffer from poor mechanical strength and screen a significant portion of the electric field from entering the flow channel. In this talk, we demonstrate a new type of conductive PDMS microfluidic FFE technique. Our approach utilizes integrated metal electrodes, directly fabricated against conductive PDMS membranes and integrated into the microchannel sidewalls. The conductive membranes are fabricated using carbon black (CB) doped PDMS, and separate the electrophoresis channel from the gallium metal electrodes. These novel CB membranes are capable of preventing electrode contamination from electrolysis, while still providing sufficient electric field for FFE. The membranes also possess high mechanical strength and do not suffer from membrane break down caused by kilovolt/meter electric field strengths. We demonstrate the capabilities of these electrode-membrane systems for on chip microfluidic sample separation and enrichment using nanoparticle based FFE-isotachophoresis and FFE-zone electrophoresis.

415035 Discussion on Nontraditional Methods of Addressing of Liquid Droplets By DC Electric Field

Abstract:

Addressing of discrete volumes of fluids in microfluidic structures to a particular detection/reaction spot is an important task in many applications. If properly imposed, DC electric field can induce electrophoretic transport to efficiently control a droplet trajectory. It is well known that water droplets dispersed in a dielectric fluid can be charged due to a faradaic reaction at a microelectrode surface. Unfortunately, there are many other two- or multiphase liquid systems, e.g. oil droplets dispersed in water phase or two immiscible water phases (based e.g. on polyethyleneglycol), in which charging cannot be carried out by this simple method. Here we will discuss three alternative techniques that could be employed in liquid addressing:

i) We have observed that electrophoretic addressing of droplets is possible in complex systems that consist of two immiscible water phases. Surface electric charge is spontaneously formed at the interface. We will show experimental arrangement, in which oriented motion of droplets was observed. Dependences of migration velocity on applied electric field will be presented.

ii) Another mechanism is related to water in oil systems. Instead of the faradaic current injection from an external electric circuit, passive charging can be carried out to get a charged droplet. The passive charging is based on the use of an electrode from a material that spontaneously dissolves in a water solution. We will critically evaluate this mechanism in the view of our experimental results.

iii) Oil in water droplets can be manipulated by DC electric field when an ionic surfactant that adsorbs to the water-oil interface is used. Electric field polarity and properly selected ionic surfactant can lead to oriented transport of oil droplets. Preliminary results that we have achieved will also be presented.

430476 The Response of an Anisotropic Colloid to a Nearby DC Electrode

Abstract:

For nearly thirty years, the directed assembly of colloidal particles proximate to an electrode has been studied in an effort to design a bottom-up process for the synthesis of two-dimensional colloidal crystals. More recently, work in this area has shifted from studying isotropic particles (with directionally independent properties) to studying the response of anisotropic particles (with directionally dependent properties). In principle, colloidal assemblies fabricated from anisotropic particles will have more complex microstructure than those fabricated from isotropic particles, which tend to form only close-packed arrays. This presentation will focus on the specific case of a geometrically anisotropic dielectric particle with non-uniform zeta potential responding to a nearby *dc* electrode. A numerical solution was obtained for the electric and flow fields in the proximity of the particle. Equilibrium charge electroosmotic (ECEO) flow is generated along the surface of the particle at these conditions, with the strength and direction of the flow depending on the magnitude and sign of the particle's zeta potential. Results from the numerical solution indicated that the ECEO flow field and the accompanying interparticle interactions were anisotropic. In addition, a net lateral force acted on the particle perpendicular to the *dc* electric field, which is very similar to induced charge phenomena experienced by a metallo-dielectric Janus particle in response to an electric field. The origin of this force is the asymmetry of the ECEO flow field, which can be tuned by tailoring the shape and surface chemistry of the particle.

Liquid Crystal Enabled Electrokinetics

Oleg D. Lavrentovich

Chenhui Peng
Yubin Guo
Dr. Sergij Shiyanovskii
Dr. Qihuo Wei

Liquid Crystal Institute
Kent State University

Collaborators:
Jorge Vinals, U Minnesota
Christopher Conklin

Support: NSF DMR and DMS

2015 AIChE Annual Meeting, Salt Lake City, UT
Session: **Soft Matter Electrokinetics: Particles, Drops and Bubbles**
November 11, 2015

Crystals, liquid crystals, liquids

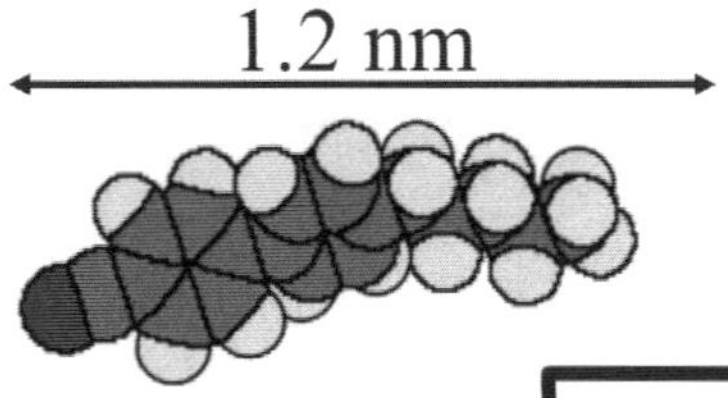

pentyl-cyanobiphenyl (5CB):
room temperature nematic, a fluid formed by elongated molecules

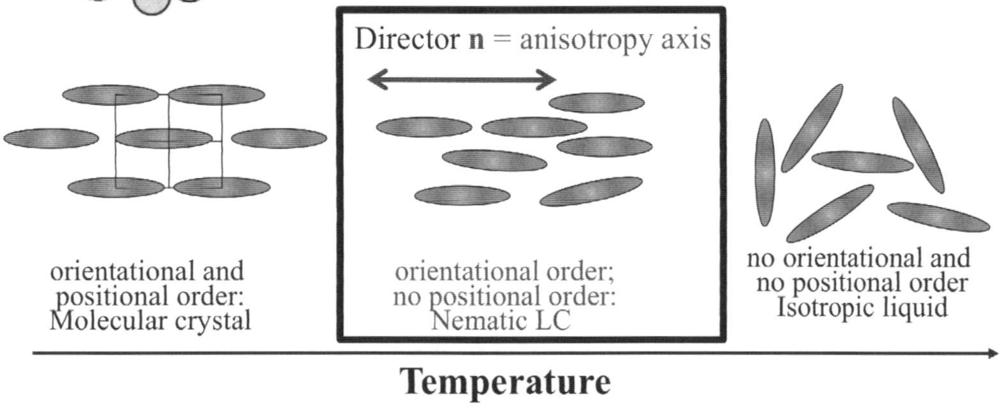

2

Two things to remember about LCs to understand LC-enabled electrokinetics

❑ Anisotropy of conductivity

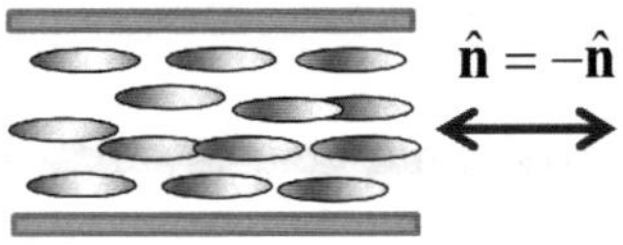

anisotropy of ionic conductivity, $\sigma_{\parallel} > \sigma_{\perp}$
(LC used in our experiments is modified to have zero dielectric anisotropy)

$n_{ions} \sim 10^9/\mathrm{m}^3$

❑ Alignment of molecular orientation: Uniform or Patterned

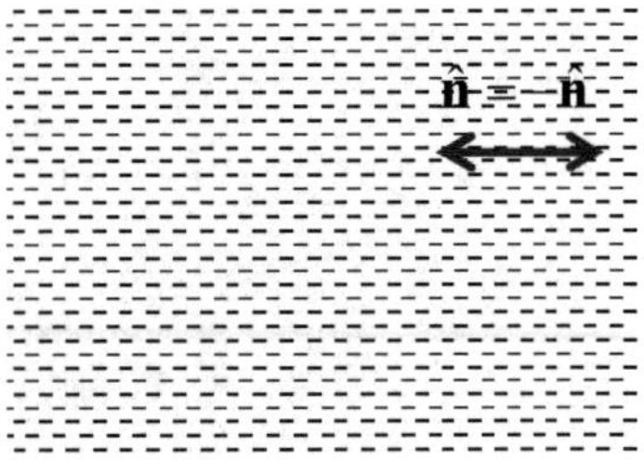

Uniform
(in your laptop and TV screens)

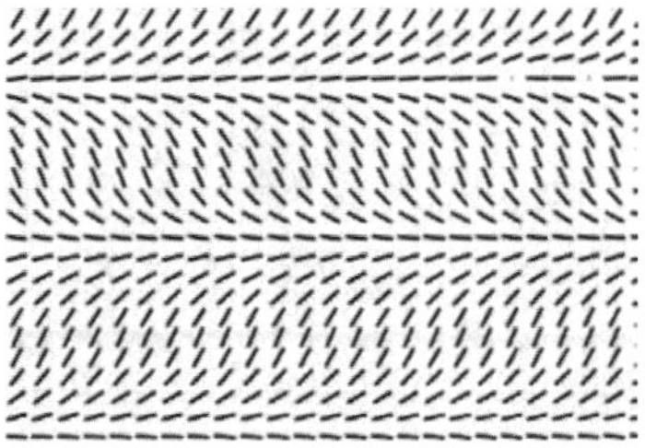

This work: molecular orientation at the substrates is patterned by photo-alignment

Distorted LC separates charges in presence of E field

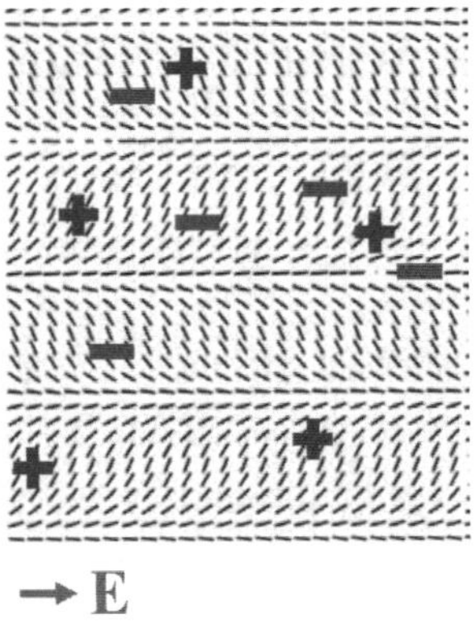

C. Peng et al, *Phys Rev E* **92**, 052502 (2015)

Distorted LC separates charges in presence of E field

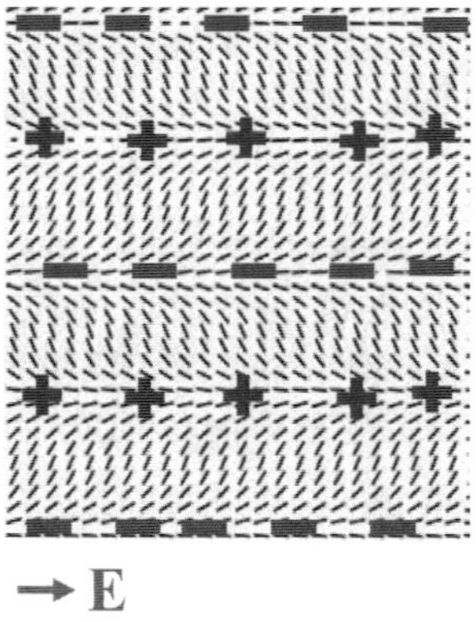

No permanent EDLs, no polarizable (metal) surfaces are needed to separate charges

C. Peng et al, *Phys Rev E* **92**, 052502 (2015)

Distorted LC separates charges in presence of E field

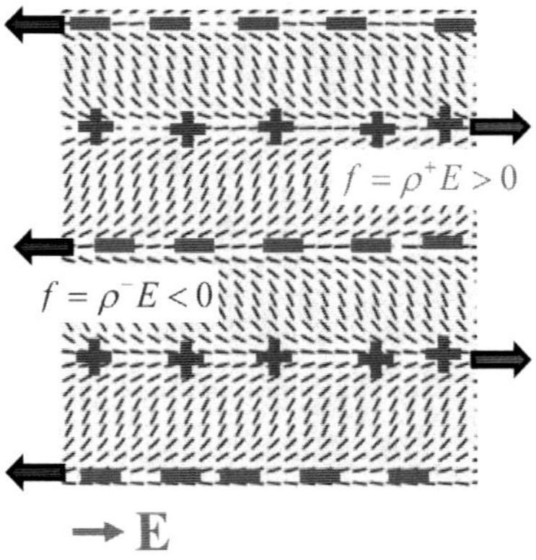

No permanent EDLs, no polarizable (metal) surfaces are needed to separate charges

C. Peng et al, *Phys Rev E* **92**, 052502 (2015)

Distorted LC separates charges in presence of E field

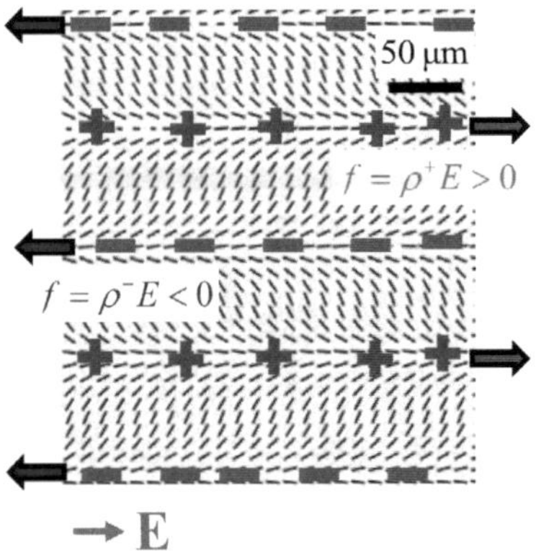

No permanent EDLs, no polarizable (metal) surfaces are needed to separate charges

C. Peng et al, *Phys Rev E* **92**, 052502 (2015)

Distorted LC creates electro-osmosis in presence of E field

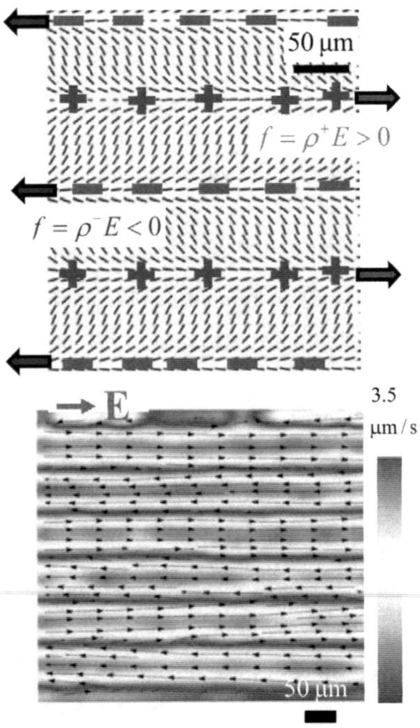

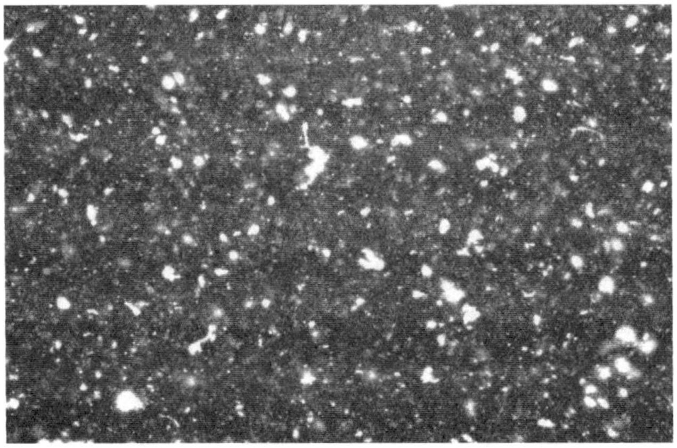

$h = 50$ μm $E = 40$ mV/μm; 5 Hz

C. Peng et al, *Phys Rev E* **92**, 052502 (2015)

Distorted LC creates electro-osmosis in presence of E field

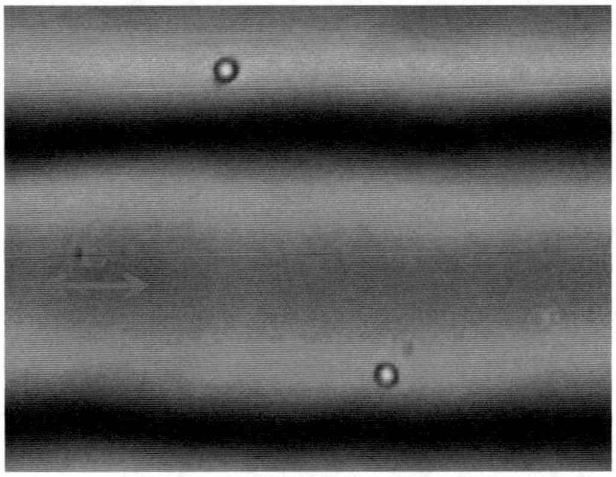

LC-enabled transport of polystyrene spheres

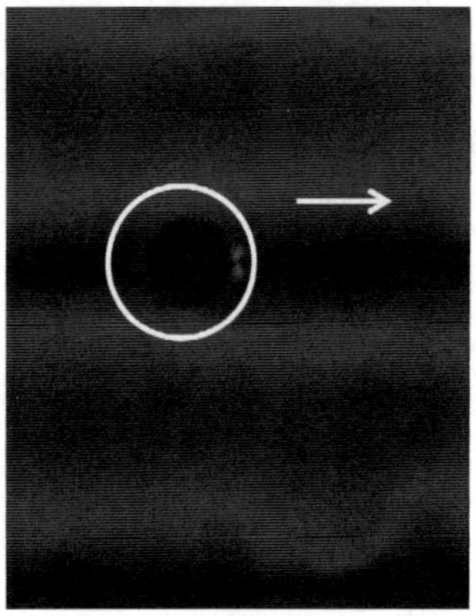

LC-enabled transport of air bubble

$h = 50\ \mu m \quad E = 40 mV/\mu m \quad f = 5\ Hz$

C. Peng et al, *Phys Rev E* **92**, 052502 (2015)

Disclination pair as an electro-osmotic pump

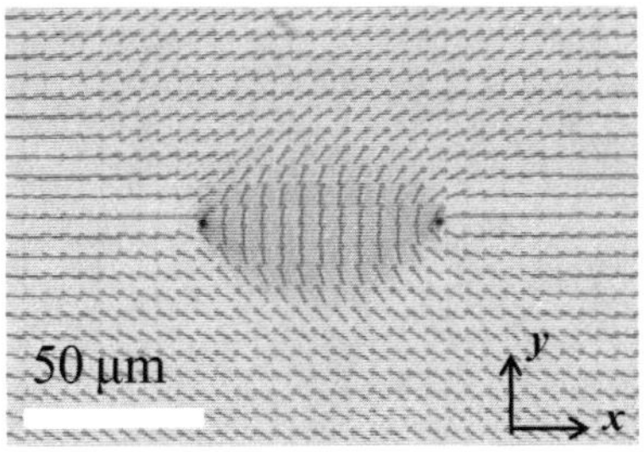

Patterned pair of (-½, ½) disclinations; dipolar symmetry;

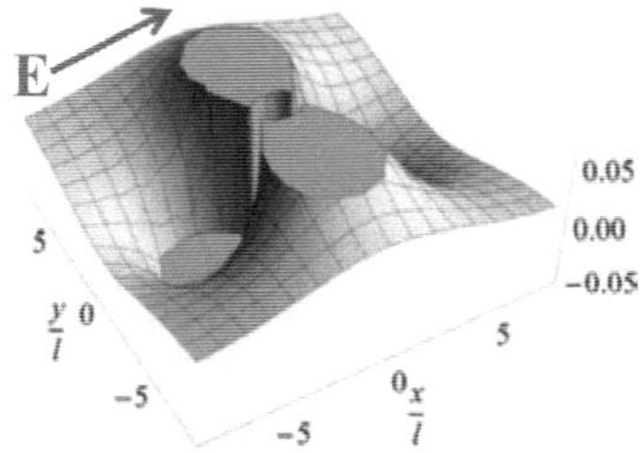

Field-separated charges

Electro-osmotic pumping: $E = 40$ mV/μm; 5 Hz

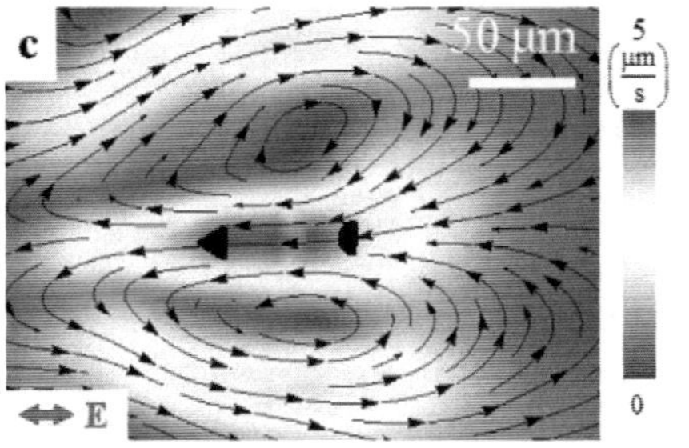

C. Peng et al, *Phys Rev E* **92**, 052502 (2015)

Disclination pair as an electro-osmotic collector of water

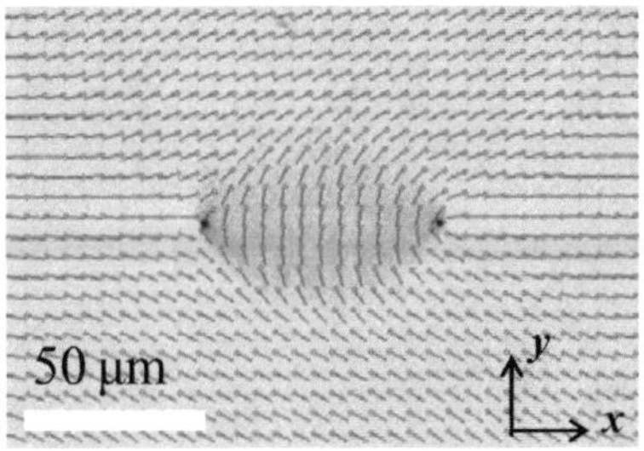

Electro-osmotic pumping: $E = 40$ mV/µm; 5 Hz

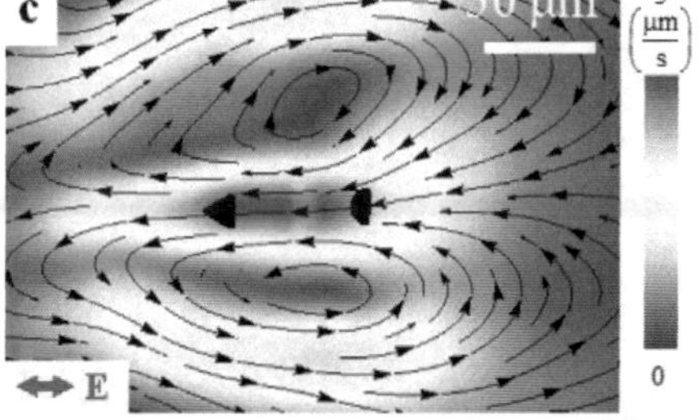

Water droplet moves towards the ½ disclination and coalesces with another droplet there

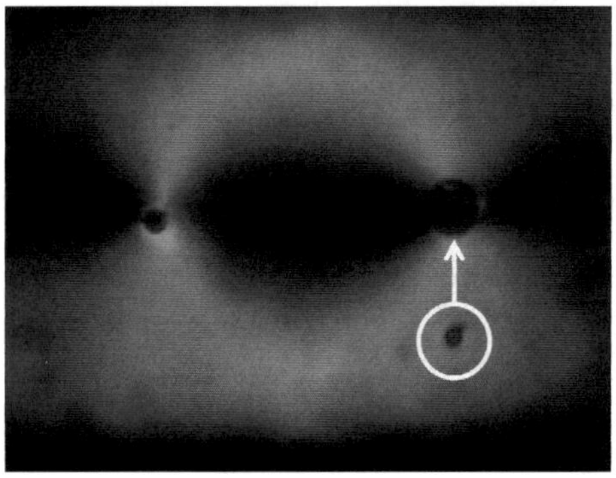

Director distortions can be created simply by placing a colloid in the chamber; LC-enabled electro-osmosis and pumping

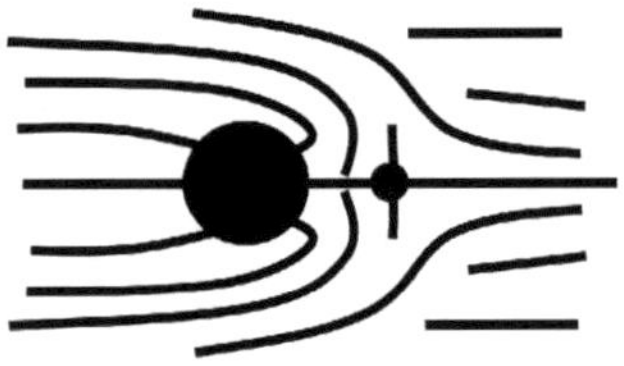

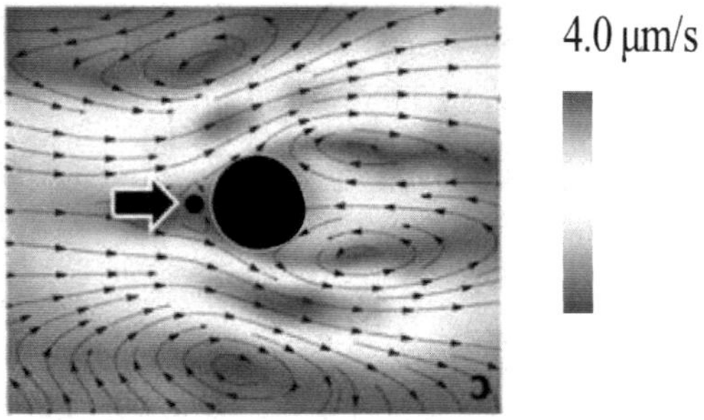

Glass sphere creates dipolar distortions of the director; in presence of E, LC is pumped along the horizontal axis

$h = 60$ µm $\quad 2a = 50$ µm $\quad E = 40$ mV/µm $\quad f = 5$ Hz $\quad \Delta\varepsilon < 0.001$

Lazo, Peng, Shiyanovskii, ODL, *Nature Comm.* (2014)

LC-enabled electrophoresis of free glass sphere

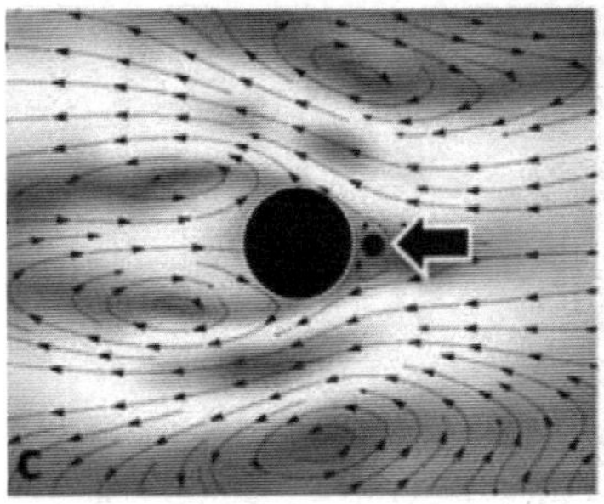

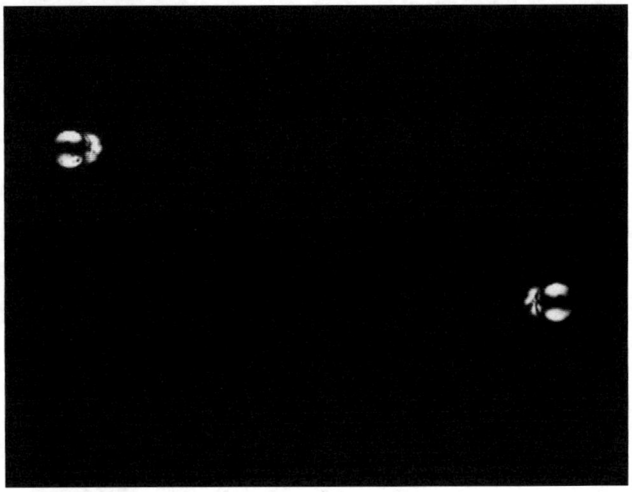

Asymmetric director, free particle: Unidirectional AC-driven electrophoresis with

$$v \propto E^2$$

Distorted LC creates electro-osmosis in presence of E field

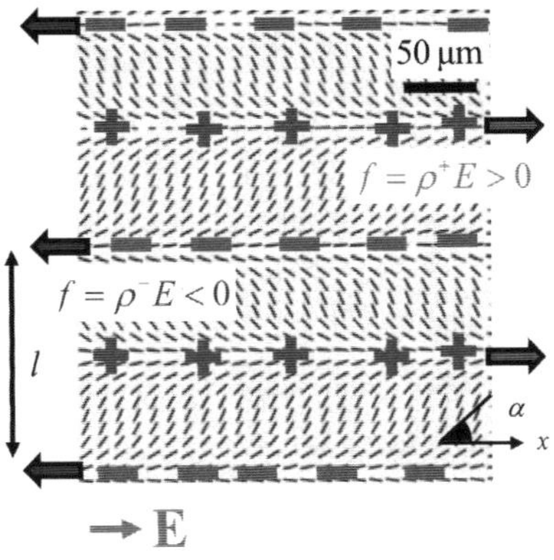

$$\mathbf{E}(t) = \mathbf{E}\,e^{-i\omega t}$$

$$\mathbf{J}(t) = \boldsymbol{\sigma}\mathbf{E}\,e^{-i\omega t} \qquad \boldsymbol{\sigma} = \sigma_{\perp}\mathbf{I} + \Delta\sigma\,\hat{\mathbf{n}}\otimes\hat{\mathbf{n}}$$

$$\mathbf{D}(t) = \varepsilon_0\boldsymbol{\varepsilon}\mathbf{E}\,e^{-i\omega t} \qquad \boldsymbol{\varepsilon} = \varepsilon_{\perp}\mathbf{I} + \Delta\varepsilon\,\hat{\mathbf{n}}\otimes\hat{\mathbf{n}}$$

$$\operatorname{div}\left(\frac{\partial\mathbf{D}}{\partial t} + \mathbf{J}\right) = \operatorname{div}\tilde{\boldsymbol{\sigma}}(\omega)\mathbf{E} = 0 \qquad \tilde{\boldsymbol{\sigma}} = \boldsymbol{\sigma} - i\omega\varepsilon_0\boldsymbol{\varepsilon}$$

$$\rho = \varepsilon_o\left(\Delta\varepsilon - \varepsilon_{\perp}\Delta\sigma/\sigma_{\perp}\right)\operatorname{div}\left(\hat{\mathbf{n}}\cdot\mathbf{E}\right)\hat{\mathbf{n}} \qquad \omega << \sigma_{\perp}/\varepsilon_0\varepsilon_{\perp}$$

$$\rho = \varepsilon_0\left(\Delta\varepsilon - \varepsilon_{\perp}\Delta\sigma/\sigma_{\perp}\right)E_0\left(\cos 2\alpha\frac{\partial\alpha}{\partial y} - \sin 2\alpha\frac{\partial\alpha}{\partial x}\right)$$

$$\rho(y) = \frac{\pi\varepsilon_0\varepsilon_{\perp}}{l}\frac{\Delta\sigma}{\sigma_{\perp}}\cos\left(\frac{2\pi y}{l}\right)E_0 \propto E_0$$

$$f = \rho E_0 \propto \varepsilon_0\varepsilon_{\perp}\Delta\sigma E_0^2/(l\sigma_{\perp})$$

$$f_{visc} \propto \eta u/l^2$$

Velocities in LC-enabled electrokinetics:

$$u \approx \varepsilon_0\varepsilon_{\perp}\Delta\sigma l E_0^2/(\eta\sigma_{\perp}) \propto \Delta\sigma, l, E_0^2$$

C. Peng et al, *Phys Rev E* **92**, 052502 (2015)

Electro-osmotic velocity around SiO_2 spheres

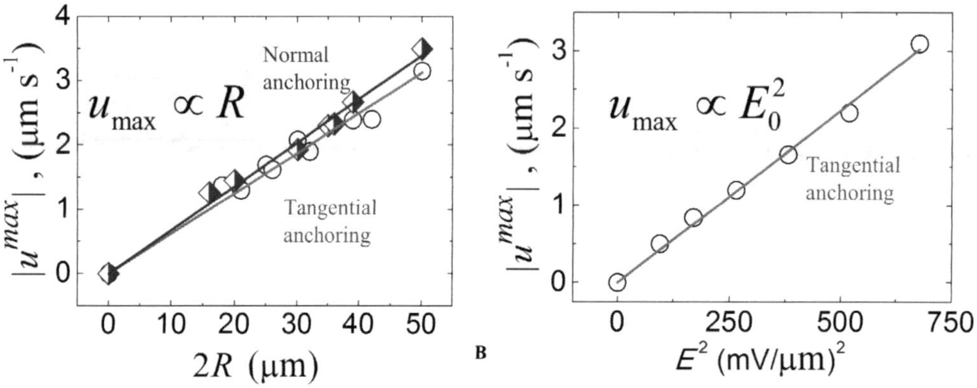

$$u \approx \varepsilon_0 \varepsilon_\perp \Delta\sigma R E_0^2 / (\eta\sigma_\perp) \propto R, E_0^2$$

Lazo, Peng, Shiyanovskii, ODL, *Nature Comm.* (2014)

Reversal or vortices by field re-alignment

Vortices are impossible in linear electrokinetics;
In LC enabled electrokinetics, they appear naturally, following director patterns

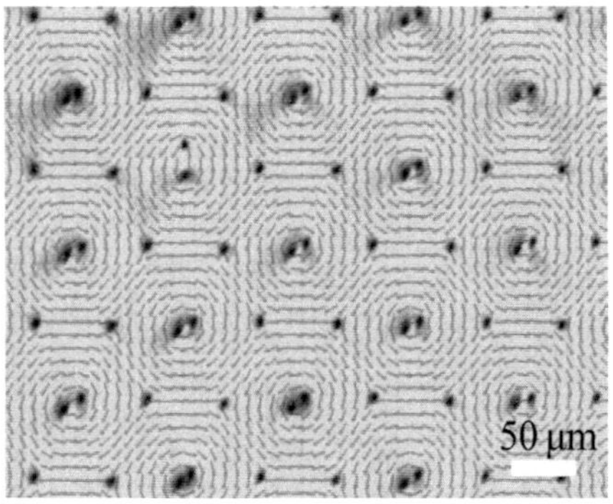

$h = 50$ μm $E = 40$mV/μm $f = 5$ Hz

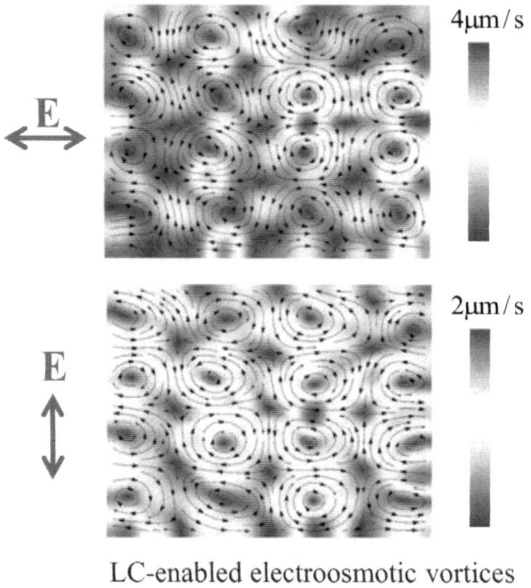

LC-enabled electroosmotic vortices reverse direction when the field is switched from (E,0) to (0,E)

Accelerated mixing

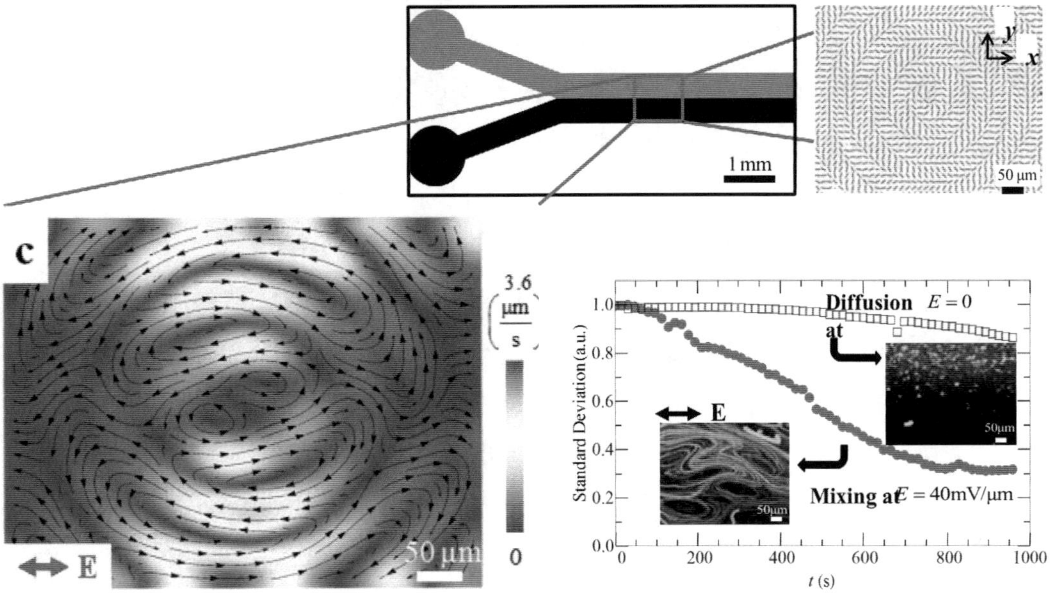

C. Peng et al, *Phys Rev E* **92**, 052502 (2015)

Accelerated mixing

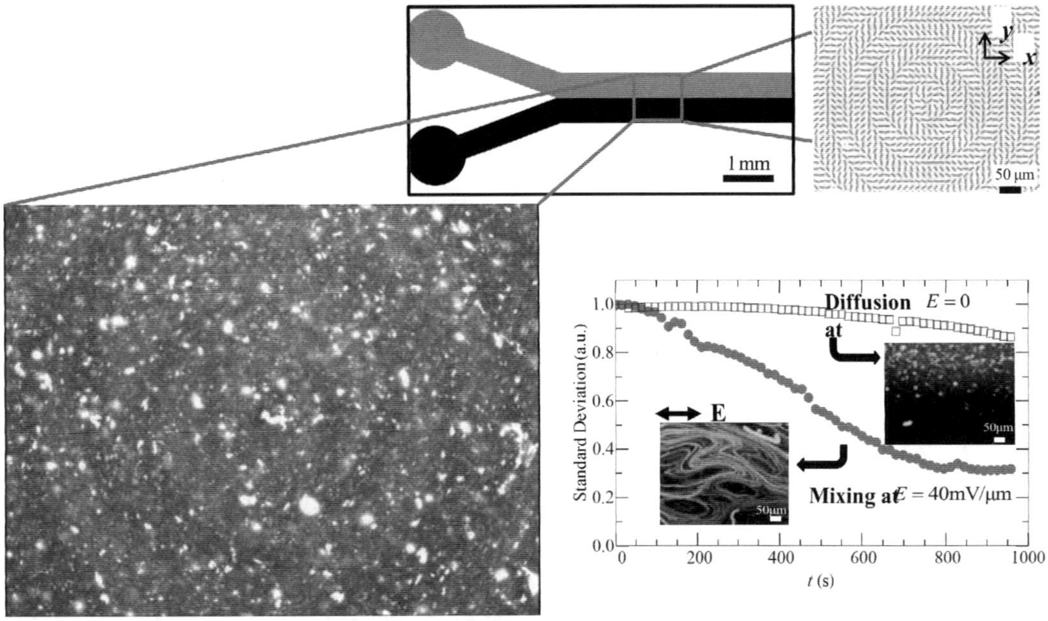

C. Peng et al, *Phys Rev E* **92**, 052502 (2015)

Conclusions

- LC-enabled electrokinetics:
  - A distorted LC enables charge separation and electrokinetics with velocities growing as E^2
  - Mechanism is rooted in anisotropy of LC electrolyte and spatial variations of molecular orientation
  - Mechanism does not require solid surfaces, neither charged (equilibrium electric double layers) nor highly polarizable (metals)
  - Patterns of LC molecular orientation enabling electrokinetics can be produced by patterned surface alignment or simply by placing colloidal particles in the chamber

Pullers and pushers

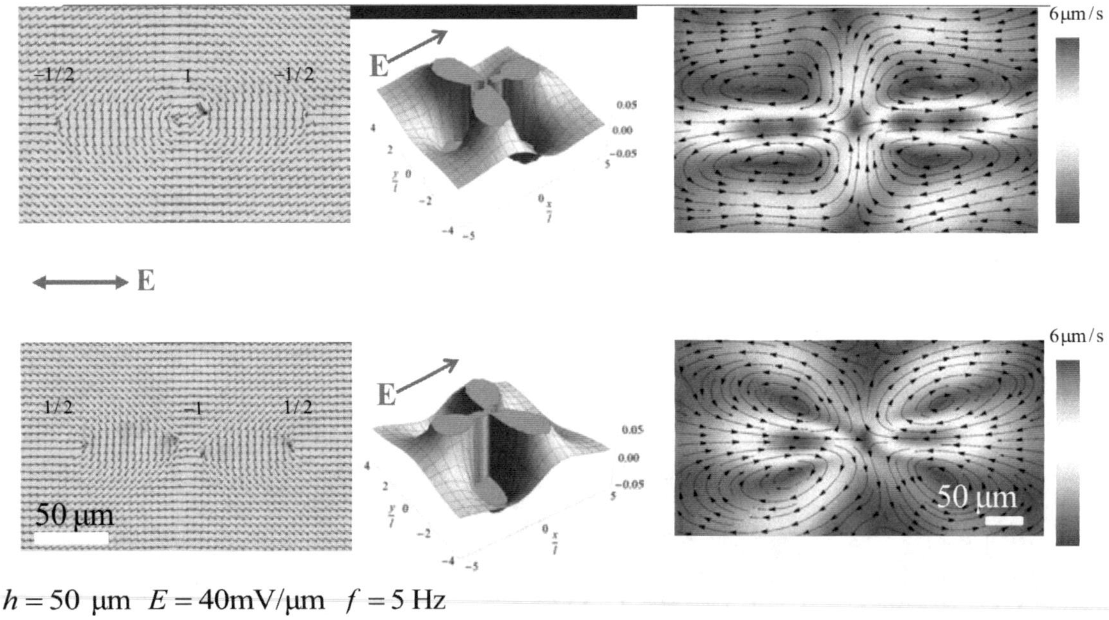

$h = 50$ μm $E = 40$mV/μm $f = 5$ Hz

C. Peng et al, *Phys Rev E* **92**, 052502 (2015)

LCI KENT STATE UNIVERSITY

AC electrophoresis of spheres in N

Alternating current (AC) driving

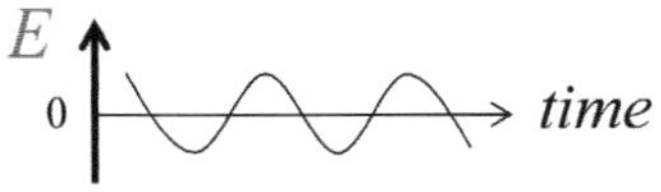

$$\langle v \rangle_{p>0} = \langle \mu E + \beta E^2 \rangle = +\langle \beta E^2 \rangle$$

$$\langle v \rangle_{p<0} = \langle \mu E - \beta E^2 \rangle = -\langle \beta E^2 \rangle$$

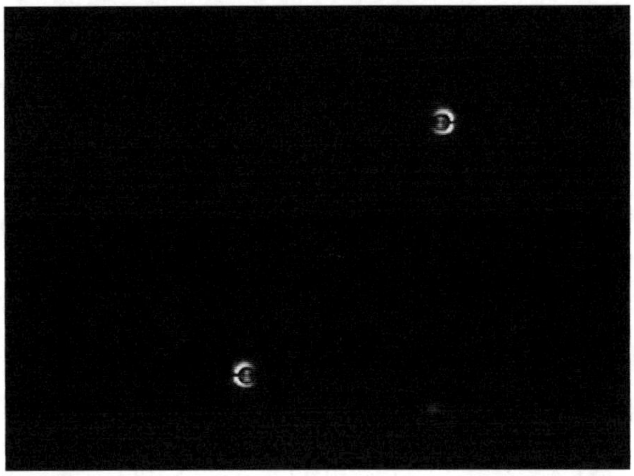

ODL, Lazo, Pishnyak, *Nature* **467**, 947 (2010)

Velocity field around SiO_2 spheres in LC

$$|u| \approx \frac{\varepsilon\varepsilon_0}{\eta}\frac{\Delta\sigma}{\sigma} R E_x^2$$

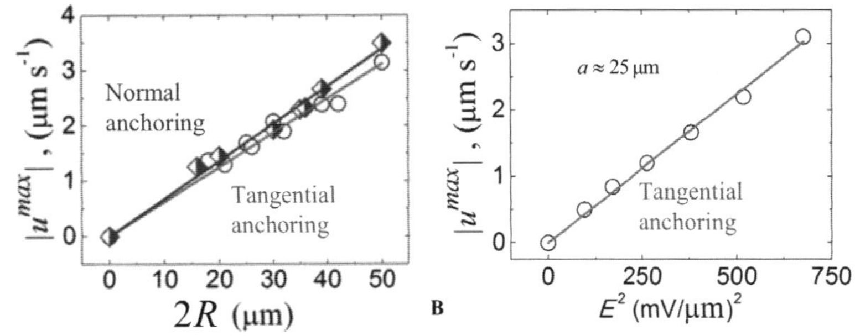

$$\frac{|u^{LC}|}{|u_{water}|} \approx \frac{\varepsilon_{LC}\eta_{water}R}{\varepsilon_{water}\eta_{LC}\lambda_D} \sim 10^2 - 10^3$$

LC-enabled electrokinetics is more efficient than linear electrophoresis in water

Lazo, Peng, Shiyanovskii, ODL, *Nature Comm.* (2014)

LCI KENT STATE UNIVERSITY

439750 Isotachophoretic Preconcentration on Paper-Based Microfluidic Devices

Abstract:

Paper substrates have been widely used to construct point-of-care lateral flow immunoassay diagnostic devices and more recently in microfluidic paper analytical devices. Paper based microfluidic devices are robust and relatively simple to operate, compared to channel microfluidic devices, which is perhaps their greatest advantage and the reason devices such as lateral flow immunoassays have reached a high level of commercial success. However, paper devices may not be well suited for integrated sample preparation, such as sample extraction and preconcentration, which may be required in complex samples with low analyte concentrations. We will present our work on integration of isotachophoresis (ITP) onto nitrocellulose-based paper microfluidic devices with the goal to improve the limit of detection of immunoassays. ITP is a powerful electrokinetic technique that has been widely used in traditional lab-on-a-chip devices to simultaneously separate and concentrate variety of ionic compounds. ITP takes advantage of a simple electrolyte system, high electrophoretic mobility leading and low electrophoretic mobility trailing electrolytes, and a constant electric field. We show that ITP on nitrocellulose is capable of up to a 1000 fold increase in initial sample concentration and up to 60% extraction from 100 µL samples and more than 80% extraction from smaller sample volumes. ITP can improve the analytical sensitivity of immunoassays by significantly increasing the concentration of the target at the test zone resulting in a dramatic increase in the surface reaction rate and equilibrium binding. We apply ITP to a Group-A Streptococcal (bacteria that causes Strep throat) immunoassay diagnostic and demonstrate a two order of magnitude improvement of the limit of detection over commercially available rapid immunoassay based lateral flow diagnostic tests.

439319 Electrokinetically Driven Microfluidic Analyzers

Abstract:

We are making micromachined systems that combine several analysis processes, with the objective of measuring disease-linked biomolecules. We photopolymerize porous polymer monoliths within microfluidic channels for solid-phase extraction involving either immunoaffinity or reversed-phase interactions. Then, purified, concentrated target compounds are analyzed through microchip electrophoresis. This approach was initially demonstrated with electrically driven extraction and electrophoretic analysis of multiple proteins related to cancer [1]. We have also implemented reversed-phase monoliths for electrokinetic on-chip extraction and fluorescence labeling of proteins [2]. Recently, we developed electrophoresis microdevices that have integrated pneumatic valves and pumps for automated pressure-based injection of precise solution volumes. We have also demonstrated field-amplified stacking and injection of non-aqueous samples in these microchips. Our present focus entails the application of these integrated microfluidic systems in studying a panel of preterm birth biomarkers. These microdevices should be useful in the analysis of small-volume biological samples.

References

[1] Yang, W.; Yu, M.; Sun, X.; Woolley, A.T. *Lab Chip* **10**, 2527–2533 (2010).
[2] Yang, R.; Pagaduan, J.V.; Yu, M.; Woolley, A.T. *Anal. Bioanal. Chem.* **407**, 737-747 (2015).

439318 Microchannel Dispersion during Isotachophoresis

Abstract:

The analytical expression for Taylor dispersion, which can be applied to isocratic capillary zone electrophoresis with a pressure-driven flow component, is often applied to zone electrophoresis in microchannels and to non-isocratic electrophoretic methods such as isoelectric focusing and isotachophoresis. This paper will discuss some of the various sources of dispersion that can occur in non-isocratic microchannel electrophoresis and provides an estimate of their impact on device resolution for the case of isotachophoresis with a laminar counterflow.

AUTHOR INDEX

AUTHOR INDEX

AIChE
120 Wall Street, FL 23
New York, NY 10005-4020

ISBN 978-1-5108-1843-9

9 781510 818439

CURRAN ASSOCIATES INC.
proceedings.com